AF865592

Nocturnal Animals - Adapted or Inherited Plasticity

Edited by Heimo Mikkola

Published in London, United Kingdom

Nocturnal Animals - Adapted or Inherited Plasticity
http://dx.doi.org/10.5772/intechopen.1011236
Edited by Heimo Mikkola

Contributors
Craig T. Perdue, Dwight P. Wynne, Heimo Mikkola, Jia Wu, Jinpei Wang, Manel Khammassi, Theodore H. Fleming

© The Editor(s) and the Author(s) 2026

The rights of the editor(s) and the author(s) have been asserted in accordance with the Copyright, Designs and Patents Act 1988. All rights to the book as a whole are reserved by INTECHOPEN LIMITED. The book as a whole (compilation) cannot be reproduced, distributed or used for commercial or non-commercial purposes without INTECHOPEN LIMITED's written permission. Enquiries concerning the use of the book should be directed to INTECHOPEN LIMITED rights and permissions department (permissions@intechopen.com).

Violations are liable to prosecution under the governing Copyright Law.

Individual chapters of this publication are distributed under the terms of the Creative Commons Attribution 4.0 License which permits commercial use, distribution and reproduction of the individual chapters, provided the original author(s) and source publication are appropriately acknowledged. If so indicated, certain images may not be included under the Creative Commons license. In such cases users will need to obtain permission from the license holder to reproduce the material. More details and guidelines concerning content reuse and adaptation can be found at http://www.intechopen.com/copyright-policy.html.

Notice

Statements and opinions expressed in the chapters are these of the individual contributors and not necessarily those of the editors or publisher. No responsibility is accepted for the accuracy of information contained in the published chapters. The publisher assumes no responsibility for any damage or injury to persons or property arising out of the use of any materials, instructions, methods or ideas contained in the book.

First published in London, United Kingdom, 2026 by IntechOpen
IntechOpen is the global imprint of INTECHOPEN LIMITED, registered in England and Wales, registration number: 11086078, 167-169 Great Portland Street, London, W1W 5PF, United Kingdom

For EU product safety concerns: IN TECH d.o.o., Prolaz Marije Krucifikse Kozulić 3, 51000 Rijeka, Croatia, info@intechopen.com or visit our website at intechopen.com.

British Library Cataloguing-in-Publication Data
A catalogue record for this book is available from the British Library

Nocturnal Animals - Adapted or Inherited Plasticity
Edited by Heimo Mikkola
p. cm.
Print ISBN 978-1-80631-235-1
Online ISBN 978-1-80631-234-4
eBook (PDF) ISBN 978-1-80631-236-8

If disposing of this product, please recycle the paper responsibly.

IntechOpen
intechopen.com

Built by scientists, for scientists

Explore all IntechOpen books

Meet the editor

Dr. Heimo Mikkola, Ph.D. in Applied Zoology and Limnology, is an Adjunct Professor at Eastern Finland University. He had a major career with the Food and Agriculture Organization of the United Nations in Africa, Central Asia, and South America. He has authored over 280 papers and books on owls and raptors and was nominated as 'Champion of Owls' in Houston, Minnesota, USA, in 2014. After retiring from the United Nations, he has edited 15 books for IntechOpen; this is book number 16 and the first on Nocturnal Animals.

Contents

Preface

One of the greatest ecological and social challenges of the 21st century is climate change. Opinion polls show that the increasing threat of climate change is at the top of public concern. The reluctance of wealthy industrial nations to reduce their reliance on fossil fuels only increases geopolitical challenges [1]. Biodiversity and climate are interconnected, and therefore, climate change forms a global threat to wildlife and ecosystems. Human activities are the primary driver of these climatic changes. So-called anthropogenic activities increase the levels of greenhouse gases (such as carbon dioxide, methane, nitrous oxides, and chlorofluorocarbons), which cause the greenhouse effect, leading to rising atmospheric temperatures and shifting thermal optima towards higher altitudes and higher latitudes. This means long-term change in weather patterns, including temperature, precipitation, and storm events [2]. The impacts of climate change on biodiversity are ongoing and are likely to result in the loss of numerous nocturnal animals, habitat fragmentation, and phenological change. Even a slight temperature change will alter species distribution and numbers, affecting behavior, reproduction, migration, and foraging. Therefore, it is critical to address climate change for biodiversity conservation and ecosystem management [3], as this book does.

Because of their vast geographical range and the incredible diversity of habitats, many nocturnal animal species are negatively impacted by one or multiple events caused by climate change: wildfires, unprecedented levels of precipitation and flooding, rising temperatures, drought, ice and snow melt, storm events, and rising sea levels [4].

Here, we examine the global threats posed by climate change to social insects and scorpions. Global warming is expected to intensify thermal extremes and aridity across many scorpion habitats. Rising temperatures, more frequent heatwaves, and prolonged droughts may exceed the tolerance limits of many scorpions and social insects, particularly for species adapted to narrow thermal niches. Luckily, nocturnality and behavioral plasticity may buffer short-term impacts; sustained shifts in temperature and humidity could alter activity windows, prey availability, and reproductive success [5].

However, it is also important to remember that animal responses to global warming may be species-specific, and not all species exhibit changing phenology despite regional climatic changes [6]. Therefore, it is challenging to separate the impacts of climate change on animal biodiversity from those of other global change drivers such as urbanization, land-use change, and increased long-distance trade [7].

Urban development significantly alters the anthropogenic challenges posed by artificial light and noise. Social insects have difficulties in maintaining their dark-friendly lifestyle.

We have a chapter on how plant-visiting bats have adapted to human life and on how a captive owl showed signs of enjoyment of human companionship.

My warmest thanks to Publishing Process Manager Laura Divic at IntechOpen for her most helpful cooperation throughout the preparation of this book.

References

[1] Dietz T, Shwom RL, Whitley CT. Climate change and society. Annual Review of Sociology. 2020;**46**(1):135–158

[2] WMO: https://wmo.int

[3] Sattar Q, Maqbool ME, Ehsan R, Akhtar S. Review on climate change and its effect on wildlife andecosystem. Open Journal of Environment Biology. 2021; **6**(1):8–14

[4] Mikkola H, Sieradzki A. The Heat is On. Global Threats to Owls from Climate Change. World Journal of Forest Research. 2025;**4**(2):1–20. DOI: 10.33140/WJFR

[5] Chown SL, Gaston KJ, Robinson D. Macrophysiology, large-scale patterns in physiological traits and their ecological implications. Functional ecology. 2004; **18**(2):159–167

[6] Vegvari Z, Bokony V, Barta Z, Kovacs G. Life history predicts advancement of avian spring migration in response to climate change. Global Change Biology. 2010;**16**(1):1–11

[7] Pautasso M. Observed impacts of climate change on terrestrial birds in Europe: an overview. Italian Journal of Zoology. 2012;**79**(2):296–314

Heimo Mikkola
Eastern Finland University,
Kuopio, Finland

Section 1

Introductory Section

Chapter 1

Introductory Chapter: Animals at Night – The Facts Behind Nocturnality

Heimo Mikkola

Abstract

The Chapter presents various ways and reasons why so many different animals have evolved to be active at night. Many of them became nocturnal to reduce the risk of being encountered by their diurnal predators. Night activities required the development of several sensorial skills, such as advanced hearing, enhanced vision, and a sharp sense of smell. Magnetic field navigation, heat detection, echo- and electrolocation were additional skills some animal groups developed during evolution. As always, the negative human impacts affected the animals, causing them to avoid human contact.

Keywords: nocturnality, evolution, sensorial skills, camouflage, magnetic navigation, human impacts, climate change

1. Introduction

Nocturnal refers to an animal, person, or plant that is active during the night. Approximately 70% of mammal species are nocturnal, while only 20% are active during the day, that is, diurnal. Another 8.5% are active irregularly during the day or night (cathemeral), and the remaining 2.5% are active at dawn and dusk (crepuscular) [1]. Of the 273 owl species, similarly to mammals, 70% of the species are nocturnal, but only three per cent are diurnal, and about 22% are partly diurnal or crepuscular. The remaining five per cent are so little studied that their activity habits are unknown [2].

Humans are diurnal because the visual adaptations needed to function in the dark are missing, and this prevents us from navigating through darkness efficiently [1]. However, some people are naturally more nocturnal than others. They feel that they can concentrate better at night, which can provide uninterrupted time for focused work, decision-making, or engaging in hobbies, like late-night sports. But normally, these activities are undertaken in artificial light, not in darkness.

This chapter explores the reasons behind nocturnal behavior, uncovering several animal groups that are active at night.

2. Evolutionary background of nocturnality

Evolutionary history suggests that the ancestors of most, if not all, mammals were nocturnal [3]. These early mammals were mainly small insectivores, living in trees and active only at night [4]. This lifestyle, known as nocturnality, evolved more than 160 million years ago as a defense mechanism against reptilian predators, such as cold-blooded dinosaurs of the *Dinosauria* group dominating the world. Ancient mammals adopted night-dwelling habits as a survival strategy to avoid detection by diurnal dinosaurs [3, 5].

Even after the extinction of dinosaurs about 65 million years ago, predator and prey dynamics have remained one of the primary reasons for nocturnality. Animals became active at night to reduce the risk of encountering diurnal predators. Conversely, predators like foxes (*Vulpes*) and many owls (*Strigiformes*) are nocturnal to hunt prey that ventures out in the dark [1–3].

3. Habitat and climate factors

Arctic habitat and climate are giving its nocturnal animals special light and temperature challenges.

The borderline between diurnal and nocturnal owls is sometimes difficult to draw, as in northern latitudes, Snowy Owls (*Bubo scandiacus*) (**Figure 1**) and Northern Hawk Owls (*Surnia ulula*) (**Figure 2**) hunt during the light nights of the Arctic summer and, in winter, during the short hours of daylight [2]. In hot climates and desert-like habitats, many animals are nocturnal to avoid daytime heat. Copperhead snakes (*Agkistrodon contortrix*), Gopher snakes (*Pituophis catenifer*), Lyre snakes (*Trimorphodon*), Milk snakes (*Lampropeltis triangulum*), and

Figure 1.
Snowy owl (Bubo scandiacus) *hunts mainly at dawn and dusk, but during the Arctic summer always in perpetual daylight. Photo courtesy of Esko Rajala.*

Figure 2.
Northern hawk owl (Surnia ulula) *hunts also at night, which is like a day in the northern latitudes during the summer. Photo courtesy of Esko Rajala.*

rattlesnakes (*Crotalinae*) sleep under logs and rocks during the day and come out at night to hunt [1, 6, 7]. Kangaroo Rats (*Dipodomys*), packrats (*Neotoma*), and other small rodents are common desert nocturnals. Foxes, Bobcats (*Lynx rufus*), and Ringtails (*Bassariscus astutus*) are mainly active at night, and bears *Ursidae* and javelinas *Tayassuidae* also become more nocturnal during hot summer months [6].

4. Competition avoidance

Clear competition avoidance exists between closely related *Asio* owls. Short-eared Owl (*Asio flammeus*) and Long-eared Owl (*Asio otus*) have been observed nesting as close as 30 meters without disturbing each other, even though they hunted the same kind of prey (mainly *Microtus* voles). Short-eared owls were more diurnal, and long-eared owls more nocturnal, thus neatly exploiting the habitat with little competition throughout the 24-hour cycle [8]. Smaller owls also adapt their normal circadian rhythms to avoid encounters with larger species. This was noted when Tengmalm's Owl (*Aegolius funereus*) and Eurasian Pygmy Owl (*Glaucidium passerinum*) bred in the same tree. The former was typically active at night, while the latter was active only during daylight hours [9].

Nocturnal activity of mammals also helps reduce competition. For instance, bobcats hunt during the day to avoid direct competition for food and water with nocturnal Coyotes (*Canis latrans*). One of the earliest primates, the Grey Slender Loris (*Loris lydekkerianus*), became nocturnal to avoid food and space competition. These lorises have excellent night vision and stealthy movements [1].

5. Better access to resources

Nocturnal animals would have a better chance of survival and success, assuming that at night there are fewer creatures active and therefore less competition for food and water [1]. It is known that many animals are active at night, as it most likely reduces predation risk. However, in low temperatures, nocturnal animals cannot escape the cold at night by sleeping in sheltered places. Thus, under natural conditions, nocturnal and diurnal activity patterns are subject to phenotypic plasticity, that is, temporal niche switching [10].

6. Adaptations for night life

Nocturnal animals have evolved various specialized traits to survive and thrive in darkness:

6.1 Advanced hearing

Some owls, such as the Barn Owl (*Tyto alba)*, the Great Grey Owl (*Strix nebulosa),* and the Long-eared Owl, use their unique hearing to locate and capture, even in the dark, the invisible prey hidden in thick vegetation or under deep snow [11]. Mammals like Wildcats (*Felis silvestris)*, the Fennec Fox (*Vulpes zerda)*, Leopards (*Panthera pardus)*, as well as rodents, also have highly sensitive auditory powers to help them hunt under the cover of darkness. Leopards can hear five times better than humans [1]. Sounds and hearing are very important to whales (*Odontocetes)* and dolphins (*Delphinidae),* because they practice long-distance communication even when underwater visibility is limited [12].

6.2 Bird and butterfly navigation

Nocturnal migrant birds have learned how to identify key directions, such as north and south. It has been noted in planetarium tests that birds change their orientation if the star pattern is shifted and become completely confused if more stars are dimmed [13]. There are indications that some birds, like European Robins (*Erithacus rubecula)*, navigate by the Earth's natural magnetic field [14]. However, it is still not fully understood how birds use the Earth's magnetic field and the stars to navigate, even during cloudy conditions. There are indications that, in some birds, iron-containing magnetite crystals are associated with their nervous system and sensory receptors, while in others, the response of visual pigments in the eye to electromagnetic energy forms the basis for geomagnetic orientation [15].

Like some birds, butterflies also seem to employ the Earth's magnetic field for navigation. Monarch butterflies (*Danaus plexippus)* and some moths catch a good tail wind at high altitudes to carry them toward their destination. This saves some energy on their unique 2,000-mile migration from eastern North America to a particular forest in Mexico each year. For navigation, Monarch butterflies use the angle of polarized sunlight as a celestial guide to help them fly straight southward. Researchers found that ultraviolet (UV) photoreceptors dominate the region of the butterfly's visual system that specializes in polarized light detection. When a UV interference filter was placed over the polarized light source in a flight simulator, the butterflies lost their orientation response. Furthermore, researchers discovered that the light-detecting sensors are connected to circadian clocks, which enable the butterflies to correct the sun angle by using their time-compensated sun compass [13, 16, 17].

7. Camouflage

Muted or patterned coloration helps animals blend into the shadows, aiding both hunters and prey [1]. Survival often lies not in strength or dominance, but in the quiet mastery of the shadows.

An adaptation that helps animals hide by blending into their surroundings is called camouflage. Animals have a color or pattern that makes them difficult to see against a similar background, like the Arctic Fox (*Alopex lagopus*) and the Snowy Owl blending with snow. Chameleons *(Chamaeleonidae)* and octopuses *(Octopoda)* change their color for camouflage or communication. Contrary to popular belief, chameleons only change color when in imminent danger [18].

8. Echolocation

Several animals, from whales to bats, use echolocation. More than 1,100 species of bats use a unique auditory system where they emit high-frequency sounds and interpret the echoes bouncing back off objects. The bat's larynx produces the ultrasonic sound at frequencies between 10 and 150 kHz [14]. Time delay and characteristics of the echoes tell the distance, size, shape, and texture of objects. This system allows them to navigate, hunt, and avoid predators in habitats with practically no light. By using high-resolution biologging tags, it was possible to prove that Greater Noctule Bats *(Nyctalus lasiopterus)* ascend to high altitudes, engage in long echo-guided chases with ultrasonic echolocation, and consume migrating passerines in mid-air during the hunting flight [19]. Only three bat species are known to eat any of the billions of small birds that seasonally migrate during the night. This far nocturnally migrating passerines at high altitudes have been kept as largely inaccessible prey to most predators, but maybe the future will bring new information when more biologging tags will be used in bird and mammal research [19].

Dolphins, porpoises (*Phocoenidae)*, and other toothed whales *(Odontocetes)* use a biological sonar system to find prey in the depths of the ocean where light is poor. Water is denser than air and transmits sound more effectively, providing the perfect setting for the use of echolocation. While using biological sonar, dolphins send high-frequency signals between 100 and 130 kHz [20]. Interestingly, dolphins have very few or no whiskers, with no intrinsic whisker muscles, as echolocation is instead an important sense for them [21]. Even in the most turbid parts of the Amazon, River Dolphins (*Inia geoffrensis)* can use acoustic echolocation to catch fish and navigate through a maze of submerged vegetation. Large whales can communicate over distances of several hundred kilometers in the ocean using very low-frequency sounds. More than 65 species of toothed whales use sonar [14]. It has been strongly stated many times that owls are not able to use echolocation when moving in the dark, just as firmly as it was claimed that they cannot use ultraviolet light as a cue to find prey. New research shows, however, that owls can detect UV light at night [22]. Interestingly, American Barn Owls (*Tyto furcata)* and related species emit metallic clicking sounds when flying, suggesting possible echolocation for orientation, and further investigation is perhaps needed to avoid similar incorrect generalizations to those involving ultraviolet light [2].

Humans can also learn to use echolocation by using self-generated sounds like mouth clicks, finger snaps, and cane taps to perceive their surroundings. With echolocation, exceptionally skilled blind people can even do mountain biking or identify objects without touching them [23]. Some individuals are able to distinguish textures, like metal, based solely on sound reflections. The brains of such individuals can reorganize to process the echoes, with the primary visual cortex showing increased sensitivity to sound. Practically, all humans can learn how to use echolocation, aiding mobility, independence, and well-being for people who are partially sighted or blind [23].

9. Electrolocation

Electroreception is known in many taxa of oceanic and freshwater fishes and amphibians, but not in mammals or insects [24]. Electric fish species, although a small minority of all fish, can generate electric fields either for defense or to stun their prey. The electric organ produces an electrical field in fish, allowing them to exploit habitats where vision is ineffective [25].

10. Enhanced vision

How well do nocturnal animals see in the dark? Owls and big cats have large eyes that allow more light in. Their eyes are so large that, in some small owls, they can take up to half of the skull area. Owl eyes are two times larger than the average for birds of the same weight. The eyeballs of some large owls are more sizable than the human eye. Even in the medium-sized Tawny Owl (*Strix aluco*), the overall length of the eye exceeds that of a human [2].

A structure called the bright tapestry (tapetum lucidum) behind their retina reflects light to enhance vision [26]. In Tigers (*Panthera tigris*), the bright tapestry or iridescent layer is greenish, but in Reindeer (*Rangifer tarandus*), the color changes seasonally (golden in summer and deep blue in winter), allowing them to better avoid predators, such as Wolverines (*Gulo gulo*), Wolves (*Canis lupus*), and Golden Eagles (*Aquila chrysaetos*), in low-light winter conditions but at the price of blurrier vision [27].

The eyes of nocturnal animals are optimized for night vision, detecting movement, and finding food in the dark. Therefore, they have more rods than cones in their eyes, and often they cannot see colors in the dark. The human eye contains 200,000 rods per square millimeter, while owls have up to a million rods per square millimeter [14]. Owls have excellent nocturnal vision without the need for a tapetum [11]. However, it is always difficult to prove whether an animal can perceive color. As owls have some cones in their retina, they can likely do so when the light is good. The crepuscular Little Owl (*Athene noctua*) can see at least yellow, green, and blue, but red and the darkest grey are confused. Nocturnal Tawny Owls can discriminate colors under bright light conditions. More diurnal owls, such as the Eurasian Pygmy Owl, can distinguish all colors [2].

11. Heat detection

Animals with this ability use specific heat-sensing receptors to detect the infrared heat radiation emitted by warm-blooded prey. Such animals include snakes, some fish, and mosquitoes. Vampire bats (*Desmodontidae)* and Forest Fire Beetles (*Melanophila acuminata*) also have sophisticated infrared sensors [14]. Most warm-blooded animals don't see the infrared spectrum but have other forms of thermoreceptors for protecting themselves from heat damage. To detect prey by sensing its body heat, rattlesnakes, boas (*Boidae)*, and pythons *(Pythonidae)* use heat-sensitive pit organs on their jaws. The heat images become integrated with visual ones in the snake's brain, which has been trained during daytime to interpret the heat images in relation to visual ones, thus enhancing the resolution of the thermal images. Rattlesnakes can detect temperature differences as small as 0.2°C [14]. Pit vipers *(Crotalinae)* possess one large organ on either side of the head, which helps them to sense radiant heat at wavelengths between 5 and 30 μm [28]. The pit organ is also used for thermoregulation, and all pit vipers are able to locate a cool refuge quickly, while true vipers *(Viperinae)* are not [29]. It is even speculated that the pit organ evolved as a defensive adaptation to aid in thermoregulatory decisions rather than a predatory one, or that multiple pressures contributed to the organ's development [30].

12. Sense of smell

Dogs are well known for their exceptionally sensitive olfactory systems, and their odor discriminatory power is a 100 times greater than that of humans. Canines have about 811 functional olfactory receptor genes in their noses, while humans have only 387 [31].

Many nocturnal animals, like rats and mice, also have an excellent sense of smell and communicate with other animals by leaving scent behind. A keen sense of smell is also used to find food [1, 31]. It is common to use trained dogs as animal biosensors for the detection of illicit drugs, explosives, and even human diseases. Trained rats would have some advantages in searching for landmines or other explosives, being lighter than dogs, not setting off explosives, and more agile in squeezing into tight holes. Most importantly, it is far more acceptable to send rats into a dangerous area than canines [31].

13. Sensitive whiskers

Most mammals have whiskers (mainly mystacial whiskers) and other specialized hairs that help animals locate their prey even in very low light. Hair and whiskers are similar in structure, but the latter are normally thicker and longer than other hair. Aquatic mammals have stiffer and thicker whiskers than terrestrial mammals. Much like human fingertips, whiskers are an active touch-sensing system. The number of whiskers varies from 4 in porpoises to around 300 in Walruses *(Odobenus rosmarus)* and several thousand in manatees *(Trichechidae)*. Whiskers do not just sense; they can also move forward and backward almost constantly, and that process is called "whisking." Mice and rats whisk 8 to 25 times per second, possibly the fastest of any movements animals can make [21]. Seal whiskers have such an aerodynamic shape that they have inspired the designers of wind turbine blades [32].

Do humans have whiskers? Only some vestigial muscle remnants exist just above our lips, in the moustache area, which are not functional. Species having whiskers tend to forage in dark and complex surroundings, such as underwater (e.g., seals *Phocidae* and sea lions *Otariidae*) or be nocturnal, climbing specialists (e.g., mice *Muridae* and shrews *Soricidae*) [33].

Night hunting scorpions, *Scorpiones,* have fine hairs on their legs that they use to sense vibrations from prey, such as small mammals [34].

14. Human impact on nocturnal behavior

14.1 Artificial light pollution

Artificial light at night and anthropogenic noise are expanding globally, acting as pervasive sensory pollutants that can disrupt wildlife behavior and reproduction. Some studies suggest that owls avoid areas with high artificial light levels [35–37]. In urban areas, where it is never truly dark, artificial lights interfere with animals' natural behavior. This is particularly harmful to migratory birds and sea turtles that navigate using starlight [1].

14.2 Climate change

As the Earth warms, some animals may shift their active hours, becoming more nocturnal or crepuscular. There is a staggering 86% chance that global average temperatures will exceed 1.5°C above preindustrial levels in at least one of the next five years, and a 1% chance of one of those years exceeding 2°C of warming [35]. Such changes can disrupt ecosystems, and negatively affect interspecific relationships and the health of animal habitats [36].

14.3 Increased human disturbance

Human presence and its related impacts on mammal diversity and activity patterns are not necessarily spatially linked to anthropogenic habitat modification. Wildlife may avoid human disturbance by adjusting both their use of space and time [37]. An animal, typically splitting its activity evenly between day and night, might increase its nocturnal activity by 68% near human disturbance. Alternatively, some prey species might enter human-dominated habitats during the night to escape nocturnal predators, spatially avoiding humans [38].

Tigers historically spread across Asia but have now lost 93% of their habitat due to human expansion [39] (**Figure 3**). Tigers are not strictly nocturnal, but to avoid conflict with humans, they tend to be most active during night and twilight hours [38]. Many of the tiger's prey, like Sambar (*Rusa unicolor)* and Indian Wild Boars (*Sus scrofa cristatus)*, are also nocturnal, and some prey, such as deer, have poor night vision, making them easier targets for the tiger. Tigers can detect faint sounds and

Figure 3.
Tiger (panthera Tigris) *tends to be most active at night to avoid conflict with humans. Photo courtesy of Ashveni and Hannu Mikkola.*

scents, which are essential for locating prey in darkness, and their night vision is estimated to be six times better than that of humans [1].

15. Conclusion

Effective and more sensitive animal sensors clearly show that humans have much to learn from nature when trying to design new infrared, radar, and acoustic sensors. Bats use echolocation to navigate, hunt, and avoid predators in habitats with practically no light. Humans can learn a skill similar to that of bats by using self-generated sounds like clicks, snaps, or taps to perceive their surroundings. This can greatly benefit blind or visually impaired people [23]. Many birds and butterflies have truly remarkable navigation capabilities that, from a weight and power standpoint, are much more efficient than any GPS navigation systems developed by humans so far [14]. Some mammals, such as dogs, mice, and rats, have a highly developed sense of smell, and they can be trained to behave like animal biosensors in many fields, not least in the medical field, for example, to assist in cancer diagnostics, etc. Through biotechnology development, it has been possible to produce genetically modified mice, which have a smelling capacity 500 times better than that of normal mice [31]. Humans have every reason to make sure that we take good care of nocturnal animals with all their exceptional sensorial skills.

Author details

Heimo Mikkola
University of Eastern Finland, Kuopio, Finland

*Address all correspondence to: heimomikkola@aol.com

IntechOpen

© 2025 The Author(s). Licensee IntechOpen. This chapter is distributed under the terms of the Creative Commons Attribution License (http://creativecommons.org/licenses/by/4.0/), which permits unrestricted use, distribution, and reproduction in any medium, provided the original work is properly cited.

References

[1] Goyal N. Why animals come out at night: The science behind nocturnality. Available from: https://medium.com/@nukul.goyal [Accessed: 2025-September-23]

[2] Mikkola H. Owls of the World. A Photographic Guide. 3rd Revised ed. London: – Bloomsbury/Christopher Helm; 2026

[3] Poudel SP, Behura SK. Relevance of the regulation of the brain-placental axis to the nocturnal bottleneck of mammals. Placenta. 2024;**155**:11–21. DOI: 10.1016/j.placenta.2024.08.001

[4] Gerkema MP, Davies WIL, Foster RG, Menaker M, Hut RA. The nocturnal bottleneck and the evolution of activity patterns in mammals. Proceedings of The Royal Society, B **280**; 20130508. DOI: 10.1098/rspb.2013.0508

[5] Hut RA, Kronfeld-Schor N, Van der Vinne V, De la Iglesia HO. In search of a temporal niche: Environmental factors. Progress in Brain Research. 2012;**199**:281–304. DOI: 10.1016/B978-0-444-59427-3.00017-4

[6] National Park Service. Animals. Available from: https://www.nps/nabr/learn/nature/animals.htm [Accessed: 2025-October-17]

[7] Shaw S. Desert Wildlife at Night. Desert USA. Available from: https://www.desertusa.com/desert-activity/nightdrive.html?srsltid... [Accessed: 2025-October-17]

[8] Mikkola H. Wood owls. In Burton JA, editor. Owls of the World. Peter Lowe - Eurobook Ltd. London; 1973. p. 116–146

[9] Mikkola H. Owls of Finland. Suomen Luonto. 1978;**37**:313–321 and 344. In Finnish, summary in English

[10] Van der Vinne V, Riede SJ, Gorter JA, Eijer WG, Sellix MT, Menaker M, Daan S, Pilorz V, Hut RA. Cold and hunger induce diurnality in a nocturnal mammal. Proceedings of the National Academy of Sciences. 2014;**111** (42):15256–15260. DOI: 10.1073/pnas.1413135111

[11] Sieradzki A. Designed for Darkness: The Unique Physiology and Anatomy of Owls. In Mikkola H, editor. Owls: Clever Survivors. Vol. **1**. London: IntechOpen; 2023. p. 3–26. DOI: 10.5772/intechopen.102397

[12] Wwl A. The Sonar of Dolphins. New York, NY: Springer; 1993

[13] Milius S. Thoroughly modern migrants: Moths and butterflies—round-trip tickets not necessary. Science News. 2004;**165**:408–410

[14] Klemas VV. Remote sensing and navigation in the animal world: An overview. Sensor Review. 2013;**33**(1): 3–13. DOI: 10.1108/02602281311294298

[15] Johnsen S, Lohmann K. Magnetoreception in animals. Physics Today 2008;**61**:29–35

[16] Hardman H. How monarch butterflies are wired for navigation. EurekAlert. Available from: https://www.eurekalert.org/pub_releases/2005-05/cp-hmb042905.php [Accessed: 2025-April-2]

[17] Sauman I, Briscoe AD, Zhu H, Shi D, Froy O, Stalleicken J, Yuan Q,

Casselman A, Reppert SM. Connecting the navigational clock to sun compass input in monarch butterfly brain. Neuron. 2005;**46**:457–467

[18] Editors of Publications International. Available from: https://animals.howstuffworks.com/animal-facts/11-animals-that-use-camouflage.htm [Accessed: 2025-October-17]

[19] Stidsholt L, Tena E, Foskolos I, Nogueras J, De la Hera I, Sánchez-Navarro S, García-Mudarra JL, Ibáñez C. Greater noctule bats prey on and consume passerines in flight. Science (New York, N.Y.). 2025;**390**:178–181. DOI: 10.1126/science.adr2475

[20] Callagher KE. https://www.treehugger.com/animals-that-use-echolocation-5112674 [Accessed: 2025-October-14]

[21] Grant R. Why don't we have whiskers? Physiology News. 2022;**126**:24–27. DOI: 10.36866/pn.126.24

[22] Höglund J, Mitkus M, Olsson P, Lind O, Drews A, Bloch NI, Kelber A, Strandh M. Owls lack UV-sensitive cone opsin and red oil droplets, but see UV light at night: Retinal transcriptomes and ocular media transmittance. Vision Research. 2019;**158**:109–119. DOI: 10.1016/j.visres.2019.02.005

[23] Thaler L. Echolocation in people. Physiology News. 2020;**126**:20–23. DOI: 10.36866/pn.126.20

[24] Bullock TH, Bodznick DA, Northcutt RG. The phylogenetic distribution of electroreception: Evidence for convergent evolution of a primitive vertebrate sense modality. Brain Research Reviews. 1983;**6**:25–46

[25] Corcoran AJ, Conner WE. Bats jamming bats: Food competition through sonar interference. Science. 2014;**346**(6210):745–747. DOI: 10.1126/science.1259512

[26] Ollivier FJ, Samuelson DA, Brooks DE, Lewis PA, Kallberg ME, Komáromy AM. Comparative morphology of the tapetum lucidum (among selected species). Veterinary Ophthalmology. 2004;7(1):11–22. DOI: 10.1111/j.1463-5224.2004.00318.x

[27] Stokkan K-A, Folkow L, Dukes J, Neveu M, Hogg C, Siefken S, Dakin SC, Jeffery G. Shifting mirrors: Adaptive changes in retinal reflections to winter darkness in Arctic reindeer. Proceedings of the Royal Society, 2013, **280**: 1–9. DOI: 10.1098/rspb.2013.2451

[28] Nature.com 2010 Snake infrared detection unravelled. Available from: https://www.nature.com/news/2010/100314 [Accessed: 2025-January-20]

[29] Krochmal AR, Bakken GS, LaDuc TJ. Heat in evolution's kitchen: Evolutionary perspectives on the functions and origin of the facial pit of pitvipers (Viperidae: Crotalinae). Journal of Experimental Biology. 2004;**207**:4231–4238. DOI: 10.1242/jeb.01278

[30] Greene HW. The ecological and behavioral context for pitviper evolution. In Biology of the Pitvipers. Campbell JA, Brodie Jr. ED Selva: Texas; 1992. p. 467

[31] Oh Y, Lee Y, Heath J, Kim M. Applications of Animal Biosensors: A Review. IEEE Sensors Journal. 2015;**15**(2):637–645. DOI: 10.1109/JSEN.2014.2358261

[32] Ahlman R, et al. Wake Flow Structure of a Seal-whisker Inspired Power Turbine Blade. Orlando, FL: AJAA Scitech Forum; 2020. DOI: 10.25 14/6.2020-2020

[33] Muchlinski MN, Wible JR, Corfe I, Sullivan M, Grant RA. Good vibrations: The evolution of whisking in small mammals. The Anatomical Record. 2020;**303**(1):89–99. DOI: 10.1002/ ar.23989

[34] Arizona-Sonora Desert Museum. Nocturnal Navigators. Arizona. Available from: https//www.desertmu seum.org>edu>docs; 2008 [Accessed: 2025-September-20]

[35] WMO 2025. Available from: https://wmo.int/ [Accessed: 2025-September-22]

[36] Mikkola H, Sieradzki A. The heat is on: The global threat to owls from climate change. World Journal of Forest Research. 2025;**4**(2):01–20. DOI: 10.331 40/WJFR

[37] Li X, Hu W, Bleisch WV, Li Q, Wang H, Lu W, Jun Sun J, Zhang F, Ti B, Jiang X. Functional diversity loss and change in nocturnal behavior of mammals under anthropogenic disturbance. Conservation Biology. 2022;**36**:1–13. DOI: 10.1111/cobi.13839

[38] Gaynor KM, Hojnowski CE, Carter NH, Brashares JS. The influence of human disturbance on wildlife nocturnality. Science. 2018;**360**: 1232–1235. DOI: 10.1126/science.aar7121

[39] Goodrich J, Wibisono H, Miquelle D, Lynam AJ, Sanderson E, Chapman S, Tne G, Chanchani P, Harihar A. Panthera Tigris. The IUCN Red List of Threatened Species 2022: e.T15955A214862019. DOI:10.2305/IUCN.UK.2022-1.RLTS.T15 955A214862019.en

Section 2

Social Insects and Scorpions

Chapter 2

Architects in the Dark: Nocturnal Behavior and Evolutionary Trade-Offs in Social Insects

Jinpei Wang and Jia Wu

Abstract

While nocturnal mammals, birds, and bats often capture the spotlight, the true architects of the dark – both in number and ecological importance – are the social insects. Taking termites, ants, and honey bees as central examples, this chapter explores the multifaceted nature of nighttime activity in these insect societies. We examine the duality of termite life: from existence in perpetual nest darkness to precisely timed, crepuscular nuptial flights. We highlight the flexibility of ants, which range from strictly nocturnal foragers to opportunistic switchers, adapting their strategies to ecological niches. And we consider honey bees – a diurnal society – whose internal, ever-dark "factory" operates with circadian precision, now threatened by light pollution. Together, these examples reveal a sophisticated evolutionary trade-off between genetically anchored adaptations – such as specialized sensory systems – and behavioral plasticity in response to environmental pressures. Finally, we discuss how artificial light, noise, and climate change are disrupting this delicate balance, putting at risk the biodiversity and ecosystem functions sustained by these vital architects of the night.

Keywords: social insects, nocturnality, behavioral plasticity, evolutionary trade-off, termites, ants, honey bees, light pollution, climate change

1. Introduction

As the Sun sets, the architects begin their work.

As the last sliver of sunlight vanishes from the African savanna horizon, the interior of a towering termite mound – a structure of earth reaching meters high – enters its most constant "day." This is not a day of light, but a "social day" meticulously maintained by millions of termites, with temperature and humidity precisely regulated. Similarly, deep within the soil beneath our feet, an ant-managed "underground metropolis," a labyrinth of tunnels, conducts busy logistics and colony expeditions under the cover of night. Inside tree hollows or human-made hives, within the dark comb cells of honey bees, workers perform the precise tasks of brood care, honey ripening, and comb maintenance, guided by an exact internal rhythm [1–3].

These scenes reveal an often-overlooked ecological truth: while nocturnal vertebrates such as mammals, birds, and bats dominate our perception, it is the social insects – particularly termites, ants, and honey bees – that are the most abundant "architects" in the nocturnal world, responsible for its most profound ecological engineering [1]. They are not merely inhabitants of the dark but active builders, transforming the night into a critical niche for their survival, reproduction, and expansion.

This chapter aims to demonstrate that nocturnality in social insects is far from a simple predator-avoidance behavior; rather, it constitutes a multidimensional adaptive strategy and an evolutionary solution. Using termites, ants, and honey bees as central examples, we systematically explore the multifaceted nature of their nocturnal activity patterns: from the extreme adaptation of termites, nearly perpetually confined to their self-constructed "eternal darkness" nests, to the behavioral flexibility exhibited by ants, ranging from obligate nocturnality to opportunistic diurnal-nocturnal switching, and further to the strict spatiotemporal division of labor in honey bee societies – "diurnal foraging and nocturnal maintenance." We argue that this diversity essentially represents a sophisticated evolutionary trade-off between genetic constraints (e.g., specialized sensory morphology) and behavioral plasticity (e.g., dynamic responses to environmental cues), ultimately serving to more effectively construct, maintain, and safeguard their complex "social architecture."

Yet, this balance, forged over vast stretches of time, now faces unprecedented disruption in the Anthropocene. Light pollution corrupts ancient photoperiodic signals, noise pollution interferes with vibration-dependent communication networks, and climate change risks unraveling the delicate synchrony between their activity rhythms and environmental factors. Understanding the survival logic of these "architects of the dark" is not only to gain insight into a fascinating biological phenomenon but also an urgent necessity for assessing how anthropogenic pressures threaten critical ecosystem functions and biodiversity. Let us now delve into this hidden, bustling world of darkness to examine the wisdom of its construction, the principles of its operation, and its present fragility.

2. Why are social insects the ultimate "architects of the dark"?

Deep in the night, as the neon lights of human cities gradually dim, an older "civilization" quietly awakens. If one could peer beneath the Earth's surface, they would see a three-dimensional metropolis composed of hundreds of millions of tiny lives: termite fortresses silently exchanging gases, ant tunnel networks bustling with material transport, and hexagonal honeybee factories steadily operating within tree hollows. These social insects are nature's greatest "architects of the dark" [4–6].

This book focuses on the core question of "why animals are nocturnal," and social insects provide the most profound answer: they are not merely inhabitants of the night but also its shapers. By constructing complex physical nests and intricate social structures, termites, ants, and bees transform darkness into a survival advantage. They represent three distinct architectural philosophies, and the differences in their nocturnal strategies perfectly illustrate the book's central theme – the evolutionary trade-off achieved between genetic adaptation and behavioral plasticity.

2.1 From circadian rhythms to the architect's time zone

The nocturnality of social insects is far from a simple behavioral "on/off switch" at a set time; it is a precise temporal program regulated by circadian systems [7, 8]. At its core is the individual molecular clock: a transcription-translation feedback loop composed of genes like period (per) and timeless (tim), forming a "master clock" in specific brain regions. Even in continuous darkness, this endogenous clock maintains a stable rhythm of approximately 24 hours, calibrated by external cues like light and temperature to synchronize precisely with the natural environment [9, 10].

However, in eusocial colonies, individual rhythms are always subservient to group needs. Young worker bees, focused on in-nest tasks, exhibit weak, indistinct rhythms; forager bees, needing to precisely interface with daytime resources, possess clear diurnal activity rhythms. This age polymorphism is direct evidence of social signals regulating individual clocks. At the colony level, a "social rhythm" emerges that is more than the sum of individual rhythms [11, 12]. Honeybees forage during the day and focus on brood care and comb building at night; termite nuptial flights are precisely timed to specific nights. Information flow and nest structure act as "synchronizers" and "buffers," respectively, bringing order and specialization to group activities in the dark.

Based on this core mechanism, the nocturnality of social insects diverges into three strategies: evasive nocturnality in ants, actively avoiding daytime heat and predators; specialized nocturnality in army ants, using programmed nocturnal raiding to create a survival advantage; and opportunistic nocturnality in termite nuptial flights, precisely matching internal and external conditions to launch "colonization plans." These three strategies are vivid footnotes in the trade-off game between genetic adaptation and behavioral plasticity [13,14].

2.2 Decoding collective intelligence and extreme adaptation

The value of studying these "architects of the dark" extends far beyond satisfying curiosity about nature.

First, they are the ultimate models for deciphering collective intelligence. In vision-limited darkness, group collaboration must rely on more primitive, fundamental communication methods [15]. Chemical pheromones, vibrational signals, tactile interactions – how do these seemingly simple signals, through local interactions among numerous individuals, give rise to complex global order in nest-building, foraging, and defense? Ant path optimization inspired "ant colony algorithms," and the honeybee "waggle dance" became a classic case of collective decision-making, providing indispensable biological blueprints for artificial intelligence and complex systems research.

Second, they are living textbooks for studying evolutionary trade-offs. Nocturnal life is a double-edged sword. The benefits are obvious: avoiding diurnal predators, reducing the risk of dehydration from high temperatures, and exploiting unoccupied temporal niches. But the costs are equally heavy: enormous energy must be invested in developing and maintaining specialized sensory systems (e.g., dim-light vision, highly sensitive olfaction), nocturnal foraging efficiency may decrease, and activity time constraints can lead to missed opportunities. Termites, ants, and bees have taken different specialized paths in weighing these gains and losses. These differences vividly illustrate that evolution has no single optimal solution but rather explores multiple possibilities under genetic constraints (e.g., body structure) and environmental pressures.

2.3 The core issues in this chapter

Recent research has revealed a startling fact: in many ecosystems around the world, the biomass of insects active at night far exceeds that of their daytime counterparts [16, 17]. This reminds us that the sunlit world represents only half of life's story. Our three protagonists embody three distinct strategies for mastering the nocturnal realm: termites are the "complete darkness specialists," their vision highly reduced, spending their lives within a network woven from chemical pheromones; ants are the "flexible nocturnal tacticians," merging dim-light vision with chemical navigation to build efficient underground logistics systems; bees are the "precise rhythm controllers," strictly separating daytime foraging from nighttime production.

These observations lead us to the central questions of this chapter and, indeed, the entire book: confronted with the shared challenge of darkness, how have these three insects developed such different "design–build–maintain" systems? How is the balance struck between the evolutionary costs and survival benefits behind these systems? Guided by these questions, we will first enter the "construction sites" of our three architects to examine their unique blueprints. Next, we will unpack the "cost-benefit ledger" underlying each strategy, finally coming to understand the multimillion-year evolutionary contest waged in the dark.

3. Navigation masters of the dark world: The nighttime engineering of the three social insects

Following the clues about rhythmic mechanisms from the previous chapter, we now step into the exclusive construction sites of our three architects. We will dismantle the core "engineering systems" behind their nocturnal survival: sensory adaptations, navigation logic, and group collaboration mechanisms. Through distinctly different evolutionary paths, they have forged three paradigms of top-tier engineering: termites, as chemical engineers of extreme environmental control; ants, as systems engineers of multimodal integration; and bees, as precision engineers of spatiotemporal resource optimization.

3.1 Termites: Chemical chief engineers of the dark fortress – the pinnacle of genetic specialization

If darkness is likened to a vast construction site without blueprints, termites are undoubtedly the most dedicated "inward-focused interior masters." They have almost entirely abandoned reliance on external light, investing all their evolutionary "chips" in an internal, autonomous ecosystem governed by chemical signals [18]. A termite nest – especially the mound of the African *Macrotermes* – is no ordinary "shelter from the elements." It is a biological architectural marvel, equipped with passive temperature control, humidity regulation, and gas exchange functions. Every structure exudes a design logic of "extreme inward focus and self-sufficiency." Recent studies have revealed that the internal temperature of mature termite mounds fluctuates by no more than 2–3°C, even when external temperatures swing between 10°C and 40°C [2]. This remarkable thermal stability is achieved through a network of interconnected tunnels and chimneys that leverage diurnal temperature gradients for natural ventilation, drawing cool air in from the base and expelling hot air from the top. Such engineering ingenuity rivals human-designed climate control systems, yet it is entirely self-sustaining, requiring no external energy input.

The sensory world of termites has been fully "devisualized." The compound eyes of workers and soldiers have degenerated to the point of near invisibility. Their sole means of perceiving the world is by "sniffing" pheromones with the dense chemical receptors on their antennae and detecting subtle nest vibrations through their feet and bodies [19, 20]. This dual "chemical + vibrational" system is their core toolkit for building in the dark. Chemical signals serve as the colony's "central nervous system." Trail pheromones act like fluorescent guide arrows on a construction site; workers returning from a food or water source release them along the path. The pheromone concentration adjusts dynamically based on the route's value and freshness, rapidly locking onto the optimal path through positive feedback mechanisms.

The vibrational early-warning system functions as an "all-weather security alarm." When a nest is breached by a predator, soldiers tap the tunnel walls with their heads. Vibrational signals race along the nest walls, instantly putting every individual in the vicinity on high alert. Even more ingenious is their use of chemical signals to "plan the workforce." Studies on *Reticulitermes speratus* have shown that these vibrational signals have distinct frequency patterns – different from those produced by worker activities like tunneling – allowing termites to distinguish between genuine threats and routine nest maintenance [21]. The queen and specific caste individuals release exclusive pheromones that inhibit the development of other individuals into reproductives, precisely maintaining the ratio of workers, soldiers, and reproductives to ensure the stability of the "construction crew." Research has identified juvenile hormone (JH) as a key regulator in this process; elevated JH levels in soldiers suppress reproductive development in larvae, while queen pheromones modulate JH synthesis across the colony [22]. This chemical "labor management" ensures that the colony never has an overabundance of nonproductive individuals, maximizing efficiency in resource-poor, dark environments.

The most spectacular "external project" in a termite's life – the nuptial flight – is a precise "colonization operation" in the dark. During specific seasons, when temperature, humidity, barometric pressure, and even moonlight achieve a perfect combination, tens of thousands of winged reproductives simultaneously surge from the nest. This is no chaotic escape but a highly coordinated expedition. Choosing nocturnal nuptial flights is, in essence, an evolved "shrewd calculation" by termites: it cleverly avoids the predatory siege of diurnal birds and reptiles, leverages the higher nighttime humidity to reduce dehydration risk, and utilizes stable air currents to achieve long-distance dispersal, scouting locations for new "fortresses" and expanding territory [23]. This "colonizing expedition" operates on a strict genetic timetable, typically occurring at dusk after rain or deep into the night. At these times, temperature, humidity, pressure, and even moonlight intensity must precisely match genetically preset thresholds, ensuring synchronized emergence and collective action to maximize the probability of population continuation. For instance, *Reticulitermes chinensis* reproductive adults undergo significant biochemical changes in their flight muscles prior to nuptial flight, including increased glycogen storage and mitochondrial activity, preparing them for the energy-intensive journey [24].

Although flight timing is strictly constrained by genetics, the specific moment can still be fine-tuned according to local rainfall patterns and predation pressure, carving out a sliver of space for behavioral plasticity within the rigid genetic framework [24, 25]. The termite strategy fundamentally relies on extreme genetic specialization to construct an efficient chemical society within the confines of the nest, one that operates independently of vision. This transforms the inherent disadvantage of darkness into an absolute advantage: mastery over the internal environment and the precise execution of "colonization projects."

3.2 Ants: The multimodal systems engineers of dynamic cities – a paradigm of flexibility

Unlike the termites' strategy of "deeply cultivating the interior and going all-in," ants more closely resemble shrewd underground urban planners who double as logistics directors. They do not abandon vision; rather, they ingeniously blend visual and chemical signals to create a highly adaptable "dual-mode navigation and collaboration system." This system allows them to function effectively in both the dim, dense environments of rainforests and under the faint light of open deserts, transforming darkness into a flexibly adjustable "safe construction window."

A key secret weapon in their nocturnal toolkit is the "hardware upgrade" of their dim-light vision system. The compound eyes of many nocturnal ants (such as the Australian bull ant, *Myrmecia*) have undergone precise modifications: their ommatidia feature larger apertures and unique crystalline cone structures, enabling them to capture more dim light [26]. Some species even possess a pupil-like contraction mechanism to dynamically regulate light intake. This equipment allows them to use moonlight, starlight, and even the polarized light of the Milky Way for orientation – essentially endowing them with a built-in "celestial compass" for accurate positioning in the dark [3, 27, 28].

Their "software core competency," meanwhile, lies in their chemical path optimization algorithm. At the micropath level, ants rely on trail pheromones for navigation, but this is far from a simple scent trail. An ant returning from a food source releases pheromones; subsequent ants tend to choose paths with stronger scent concentrations. Over time, this process continuously reinforces the shortest and most efficient route, forming a collective optimal solution. This very mechanism is the biological prototype for the computer science field's "ant colony optimization algorithm" [29–32]. Even more remarkable is the ants' "elastic labor management." Worker tasks are dynamically adjusted with age: younger individuals specialize in "indoor jobs," such as brood care and cleaning within the nest, while older individuals undertake "high-risk outdoor tasks," like nocturnal foraging and defense [33]. Research has found that the structure of the mushroom bodies – the core brain region responsible for learning and memory – differs significantly among ants with different specializations. This is akin to cultivating dedicated "professional talent" specifically tailored for different "construction roles" within the colony [34, 35].

The ant strategy harbors astonishing behavioral plasticity. Instead of betting everything on a single sensory modality, they rely on the integration of multimodal information and a flexible social organization to dynamically adapt to the unpredictable nocturnal environment. They neither go "all-in" like termites nor adhere strictly to fixed routines like honeybees. Instead, they have stubbornly forged a "versatile and adaptive" path to survival in the dark.

3.3 Honeybees: Optical timing engineers of the precision factory – a masterpiece of rhythmicity

The engineering brilliance of honeybees is most strikingly revealed in their extreme partitioning of space and time, coupled with their precise mastery of material efficiency. While celebrated for their daytime foraging, the true heart of their survival – the hive – is a precision manufacturing plant that operates in perpetual darkness, maintained at a constant temperature and humidity [36]. The nocturnal nature of honeybees is expressed primarily in the meticulous maintenance of this dark factory, though some species have also cleverly exploited a twilight foraging window, maximizing the functional value of both day and night.

Within the hive, where not a trace of light penetrates, worker bees collaborate with remarkable efficiency using touch, temperature sensitivity, and chemical signals. They precisely nurture larvae, transform nectar into honey, and construct hexagonal cells from self-secreted beeswax. This architectural form achieves maximum spatial utility with minimal material – a structure recognized in nature as the "gold standard of efficiency." At night, the colony works collectively through fanning or clustering to maintain hive temperature within the precise 32–35°C range ideal for larval development, ensuring uninterrupted factory production and brood care [37, 38]. The nocturnal nature of honeybees is expressed primarily in the meticulous maintenance of this dark factory, though some species have also cleverly exploited a twilight foraging window, maximizing the functional value of both day and night. The hive's architecture is a marvel of bioengineering: hexagonal cells minimize material use while maximizing storage space, and the comb's vertical orientation facilitates efficient temperature regulation and resource transport – all designed to optimize the colony's operations in both light and dark environments.

During daylight hours, honeybees demonstrate mastery in polarized light navigation. Their compound eyes can decode the polarization patterns created by scattered sunlight, enabling accurate directional judgment even under cloudy skies. The renowned "waggle dance" functions as an abstract communication language within the illuminated hive, precisely encoding both the direction and distance to food sources [39, 40]. Yet this highly efficient system possesses a critical limitation: it cannot operate in complete darkness, which restricts large-scale nocturnal foraging. Some bee species, such as those in the Megalopta genus, have evolved enhanced low-light vision to become "crepuscular foragers." Active during twilight hours after sunset, they visit night-blooming flowers, thereby accessing a specialized resource zone with reduced competition [41, 42].

The honeybee strategy represents an exquisite division of labor governed by strict circadian rhythms: high-risk "raw material collection" occurs during daylight when sensory efficiency peaks, while high-value "production processing" takes place within the secure, controlled darkness of the hive. This strategy's success depends fundamentally on precise synchronization between their internal biological clock and external light-dark cycles, rendering it particularly vulnerable to light pollution. For example, bees exposed to nightly light may initiate foraging earlier than necessary, wasting energy on fruitless trips, or fail to coordinate night-time maintenance tasks, leading to reduced brood survival and honey production. Should these rhythms become disrupted, this precision factory faces imminent operational crisis – a vulnerability that highlights the trade-off between the strategy's efficiency and its sensitivity to environmental change.

4. The ledger of evolutionary trade-offs: Benefits and costs of nocturnal engineering

The previous chapter revealed the ingenious engineering designs of our three "architects of the dark." Yet, in evolution, there is no such thing as a free lunch. Behind each remarkable adaptation lies a meticulous "evolutionary ledger," recording the delicate balance between survival benefits and physiological costs.

4.1 Survival dividends under dark shelter

By choosing to "build" at night, social insects first secured three fundamental survival advantages – foundations upon which their entire engineering prowess could evolve.

The foremost advantage is the "safe construction window." For vulnerable groups like worker ants and bees, safety is paramount. Night functions as a vast temporal sanctuary. Their primary predators – birds, lizards, parasitic wasps – are predominantly visual hunters active by day. Shifting high-risk activities like foraging and nuptial flights to the darkness dramatically reduces colony exposure. This evolutionary strategy finds compelling evidence in global research showing nocturnal insect biomass consistently surpassing diurnal biomass – a clear testament to the survival advantage of "nocturnal predator avoidance" [43].

The core advantage is the "environmental buffer." As ectotherms, insects' metabolism depends directly on ambient temperature. Daytime heat poses dual threats: risk of overheating and accelerated moisture loss through evaporation. For termites that must maintain specific nest humidity and for delicate larvae, such conditions can prove fatal. Night's cooler temperatures and higher humidity mitigate both water loss and thermal stress, creating optimal working conditions. Moreover, natural darkness preserves their internal biological clocks, preventing artificial light from disrupting the precise physiological rhythms governing pheromone release and coordinated activity [2, 24, 25, 44].

A third strategic advantage emerges: "temporal niche specialization." In ecological systems, species reduce competition by partitioning resources across different dimensions. Nocturnality represents a masterful temporal division strategy. While diurnal bees compete fiercely for daylight-blooming flowers, crepuscular species have specialized in plants that flower at dusk or night – such as tuberoses and certain cacti. By exploiting this temporal niche, they access a veritable "blue ocean" of resources with minimal competition, securing stable nourishment [45].

4.2 The evolutionary price of adapting to darkness

To access these survival dividends, social insects have paid a steep price, primarily across three dimensions: sensory investment, performance limitations, and opportunity costs.

First is the development and maintenance cost of "specialized equipment." Perceiving the world in darkness demands specialized sensory and neural systems, requiring continuous energy expenditure. Termites, to build their "chemical internet," must synthesize complex pheromones and evolve highly sensitive chemoreceptors – sacrificing vision and forfeiting the ability to detect fast-moving threats or recognize landmarks. Ants attempt to "have their cake and eat it too," maintaining both dim-light vision and chemical systems. This results not only in more complex sensory structures but also places greater information integration demands on their brains, creating severe energy budget constraints [46]. Honeybees' "optical clockwork," while precise, remains acutely vulnerable to disruption; cloudy conditions or light pollution can disable their navigation, with this systemic fragility constituting a core cost.

Second is the "inherent ceiling" on behavioral efficiency. No matter how specialized their senses, darkness fundamentally limits performance [47]. Ants traverse more circuitous paths at night, reducing exploration efficiency. The honeybee "waggle dance" becomes useless in darkness, forcing nocturnal species to rely on less informative communication methods, thereby limiting the complexity of group coordination [48]. More crucially, most resources – like the nectar of flowering plants – remain unavailable at night. Choosing nocturnality thus means voluntarily abandoning daytime's "resource bonanza," leaving survival heavily dependent on limited nocturnal resources.

The most profound cost is opportunity cost – the sacrifice of daytime options to specialize for the night. Obligately nocturnal species like termites and certain ants have nearly completely lost the capacity to function or compete during daylight. Their physical morphology and social structures have become "custom-designed for darkness," representing an evolutionary "all-in" gamble. Should their dark environment be compromised, their adaptive advantages could vanish instantly.

4.3 The art of compromise: Not optimal, only suitable

The cases of termites, ants, and bees vividly illustrate a core principle of evolution: there exists no universally optimal solution, only locally optimal adaptive peaks shaped by genetic history, physiological constraints, and ecological pressures [49].

Termites opted for "extreme specialization," betting everything on chemical communication to achieve absolute dominance in subterranean darkness – at the cost of flexibility [50]. Ants chose "flexible balance," relying on dual-modal systems and dynamic task allocation to adapt to variable environments – at the cost of high energy expenditure [47]. Bees selected "rhythmic partitioning," maximizing efficiency across day and night phases – at the cost of heavy reliance on photoperiods [51]. Their success lies not in flawless strategies but in achieving a balance where benefits sufficiently outweigh costs within their specific evolutionary contexts, enabling colony survival and reproduction. This equilibrium within the evolutionary ledger is far more dynamic and intricate than we often imagine. In the next chapter, we will witness how human activity is now altering this ancient ledger, disrupting hard-won balances and confronting these architects of darkness with an unprecedented "accounting crisis."

5. Crisis in the dark: The Anthropocene challenge to the architects of darkness

Having explored the engineering ingenuity of social insects and unpacked the logic of their evolutionary compromises, we now face a new challenge to this multimillion-year success story, written in darkness. Humanity is reshaping Earth's lightscape and climatic rhythms at an unprecedented pace and scale. Light pollution, noise pollution, and climate change – three unseen hammers – are destabilizing the very foundations of these architects' existence, pushing their meticulously evolved balance toward the brink.

5.1 Light pollution: Hijacking the ancient celestial compass

The most immediate and pervasive threat originates from our compulsion to "illuminate the night." Artificial light at night [52], particularly from blue-rich LEDs, is casting a "perpetual twilight" across the globe. This acts as an inescapable noise signal superimposed on insects' navigational interfaces, fundamentally corrupting their "construction commands" and circadian timetables [53].

For honeybees, light pollution scrambles polarized light navigation, causing disoriented foragers to circle artificial lights until exhausted [54]. It also disrupts their biological clocks, triggering abnormal nocturnal activity and chaotic brood care, effectively paralyzing the precision hive. For termites, artificial light severely interferes with nuptial flight rhythms [25]. It can falsely trigger premature swarming, sending vulnerable reproductives into peril, or suppress swarming entirely, cutting off the vital process of dispersal and new colony foundation. Moreover, the lights attract visual predators like birds and bats, transforming these critical events into targeted slaughters that undermine the colony's future. For ants, artificial illumination disrupts their "diurnal-nocturnal work shifts" [55]. Obligate nocturnal species are forced to curtail their activity, while opportunistic foragers may misjudge safe periods, exposing themselves to daytime heat and predators, drastically slowing the expansion of their subterranean metropolises.

5.2 Noise and vibration pollution: A trembling subterranean world

Beyond light, we flood the environment with mechanical energy – the roar of traffic, industrial vibrations, and construction clamor. This creates a constant background chaos that precisely sabotages the "vibrational communication networks" essential to ants and termites.

Termites rely on vibrations traveling through nest walls to detect predator excavation. Persistent low-frequency noise drowns out these alarm signals, leaving colonies vulnerable to undetected invasion. It also interferes with pheromone communication, disrupting the internal order of construction and maintenance [56]. Ant navigation and spatial memory are highly vibration-sensitive. External vibrations can disorient foraging workers, causing them to lose their way. Intense vibrations can even cause tunnel collapses and destroy functional chambers, forcing colonies into energetically costly relocations and rebuilds [57]. While honeybees are less vibration-sensitive, severe disturbances can still incite panic among workers, leading to breakdowns in hive thermoregulation and brood care, impairing the hive's smooth operation [58].

5.3 Climate change: Undermining the physical foundations

Climate change brings more than rising temperatures – it delivers comprehensive disruptions to temperature, humidity, precipitation, and phenology. Like shaking

DOI: http://dx.doi.org/10.5772/intechopen.1015011

the very foundations of a building, it severs the critical link between insects' "construction conditions" and their optimal timing.

The passive climate control systems of termite mounds and ant nests evolved under historically stable climatic conditions [59]. Extreme heatwaves, prolonged droughts, and intense rainfall now exceed their regulatory capacity, destabilizing the delicate internal environment [4]. Honeybee foraging rhythms are tightly synchronized with plant flowering phenology. Climate-induced phenological mismatches can leave bees emerging to find their essential food sources not yet in bloom, plunging colonies into "ecological famine." Unusual rainfall can flood subterranean nests, while droughts can desiccate water sources – each posing direct, existential threats [58].

More critically, these pressures rarely act in isolation. Their synergistic effects amplify the damage. Light pollution combined with urban heat islands intensifies nocturnal heat stress, overwhelming the temperature regulation of bees and ants [58]. Insects already weakened by drought are further depleted when attracted to artificial lights for futile flights, hastening their demise. Phenological mismatches coupled with increased resource competition force nocturnal insects to shift their activity windows, bringing them into direct conflict with diurnal species and further squeezing their ecological space.

5.4 A vulnerability map of the three architects

Confronted by these crises, the three architects exhibit a gradient of vulnerability shaped by their distinct strategies. Termites – the extreme specialists – prove most fragile. Their highly specialized chemical-vibrational systems have minimal tolerance for environmental change, and their fixed, immobile nests prevent rapid relocation. Once key environmental parameters shift beyond their narrow range, adaptation becomes nearly impossible [60]. Honeybees – the precision rhythm controllers – are acutely sensitive. Light pollution and phenological mismatch strike directly at their core systems, leaving them with almost no margin for error [58]. Ants – the flexible balancers – demonstrate relative resilience, using behavioral plasticity to buffer against certain pressures [61]. Yet even their adaptability has limits; the cumulative, multidimensional stresses of the Anthropocene will eventually overwhelm their capacity to cope.

These novel pressures of the human era are rewriting the rules of the dark world at a pace far exceeding the natural evolutionary response, pushing these ancient architects toward the very edge of survival.

6. Redrawing the blueprint of darkness, rebuilding the web of life

Through a comprehensive exploration of termites, ants, and bees – three "architects of the dark" – spanning microscopic senses and macroscopic ecology, evolutionary history, and Anthropocene challenges, we have discovered that their nocturnal behavior is far more than a simple biological clock preference. It constitutes a sophisticated survival engineering system that intricately blends specialized genetic hardware with adaptable behavioral software. This final chapter synthesizes our core findings and looks ahead: in an era where humanity profoundly reshapes the planet, how can we reassess the ecological value of darkness and protect these silent builders who help maintain Earth's stability?

6.1 Three architectural wisdoms in the dark, one evolutionary logic

This exploration has revealed three key conclusions:

First, the diversity of strategy demonstrates evolutionary creativity. Faced with the common challenge of "building complex societies in darkness," nature has evolved three distinct solutions: the termite's "inner fortress" strategy, relying on extreme genetic specialization to create self-contained systems; the ant's "flexible network" approach, utilizing multimodal integration to adapt to changing environments; and the honeybee's "spatiotemporal division" method, employing rhythmic task allocation to maximize efficiency. No single strategy is superior; each represents a successful model within its specific ecological niche, highlighting the remarkable range and depth of biological adaptation [4, 9].

Second, all successful strategies are grounded in precise evolutionary trade-offs. Nocturnality offers no free advantages. Every adaptation carries a tangible cost: developing specialized senses requires energy investment, darkness inherently limits behavioral efficiency, and choosing the night means forgoing daytime resources [62–64]. The "success ledger" of termites, ants, and bees fundamentally reflects, within their respective contexts, a dynamic equilibrium where benefits – such as predator avoidance, water conservation, and niche expansion – outweigh costs, balancing genetic constraints against environmental pressures.

Third, human activity is systematically disrupting this ancient equilibrium. Light pollution, noise pollution, and climate change are not natural evolutionary pressures [59, 65]. They attack the vulnerabilities of insect strategies at a pace exceeding nature's tempo, and their synergistic effects are particularly destructive, pushing systems refined over millions of years toward the edge of collapse.

6.2 Future outlook: A three-pronged approach in science, conservation, and ethics

Based on this understanding, urgent action is needed across three interconnected domains: science, conservation practice, and ethics.

Scientifically, we must advance cross-scale integrated research, connecting genes, neurons, individuals, colonies, and ecosystems to build holistic models. Priority should be given to uncovering the neural regulatory mechanisms behind different nocturnal strategies, deepening our understanding of how individuals perceive and process environmental information to translate it into collective action, and achieving a comprehensive grasp of swarm intelligence. Concurrently, leveraging genomics and neurobiology, we should identify the "architectural genes" involved in nest building, sensory adaptation, and rhythm regulation. Clarifying the molecular basis of termite vision loss, ant behavioral plasticity, and bee circadian stability is crucial. Long-term field monitoring is essential to track changes in the "construction behaviors" of these insects under human influence – such as shifts in termite nuptial flight timing, ant nest structural adjustments, and bee rhythm fluctuations – providing a robust data foundation for conservation planning.

In conservation practice, the focus must shift from protecting individual species to safeguarding ecological processes, implementing "darkness-friendly" conservation frameworks. We should promote the "Dark Sky" initiative, reducing unnecessary artificial lighting in urban planning and adopting insect-friendly light sources (e.g., low-intensity, warm-spectrum lighting) to minimize interference with their "construction signals." Establishing climate-resilient corridors can provide suitable migration and breeding pathways for insects, buffering the impacts of climate change on their "architectural foundations." Embracing nature-based solutions, such as preserving native vegetation, offers insects continuous resources and microclimatic refuges.

Ethically, we must fundamentally rethink our relationship with darkness. Darkness is not an empty void to be conquered but an essential ecological resource – the vital "workspace" for nocturnal life. Protecting darkness means safeguarding biodiversity and ecological resilience, preserving the integrity of Earth's life-sustaining networks. Ultimately, human prosperity depends on ecological balance. Termites drive nutrient cycles, ants help control pests, and bees ensure pollination. Their decline will inevitably circle back to undermine human well-being.

7. Conclusions

For millions of years, termites, ants, and bees have been composing a silent yet profound epic in the dark. Their constructions are not merely physical nests but indispensable nodes within the web of life. Today, humanity holds the power to rewrite Earth's environmental narrative. We stand at a crossroads: will we continue as unaware "disruptors," accelerating the demise of these ancient architects, or will we choose to become mindful "stewards," preserving the dark stage essential for their continued existence?

Choosing the latter path means recognizing that true civilization lies not in illuminating every corner, but in knowing which darkness deserves preservation; not in conquering nature, but in learning to coexist with the rhythms of other lives. When we safeguard a stretch of night for these "architects of darkness," allowing them to continue planning, building, and thriving in freedom, what we ultimately protect is the very possibility for Earth's life story to endure – and, with it, the foundation for our own sustainable future on this planet.

This stewardship must begin now.

Acknowledgments

Natural Science Foundation Project of the Science and Technology Department of Shaanxi Province (2025JC-YBMS-247), and the Scientific and Technological Plan Project of Xi'an Science and Technology Bureau (24GXFW0080).

Conflict of Interest

The authors declare no conflict of interest.

Author details

Jinpei Wang* and Jia Wu
Applied Research Center for Life Sciences, Xi'an International University, Xi'an, P. R. China

*Address all correspondence to: wjplwp22@126.com

IntechOpen

© 2026 The Author(s). Licensee IntechOpen. This chapter is distributed under the terms of the Creative Commons Attribution License (http://creativecommons.org/licenses/by/4.0/), which permits unrestricted use, distribution, and reproduction in any medium, provided the original work is properly cited.

References

[1] Roeder KA, Roeder DV, Jelena B. Ant thermal tolerance: A review of methods, hypotheses, and sources of variation. Annals of the Entomological Society of America. 2021;**114**(4):459–469. DOI: 10.1093/aesa/saab018

[2] King H, Ocko S, Mahadevan L. Termite mounds harness diurnal temperature oscillations for ventilation. Proceedings of the National Academy of Sciences of the United States of America. 2015;**112** (37):11589–11593. DOI: 10.1073/pnas.1423242112

[3] Freas CA, Narenda A, Murray T, Cheng K. Polarised moonlight guides nocturnal bull ants home. elife. 2024;**13**:RP97615. DOI: 10.7554/eLife.97615

[4] Nengi Fiona Karibi-Botoye GT, Muljadi B, Demyanov V, Singh K. Termite mound architecture and climate control: A review of X-ray tomography and flow field simulation approaches. Journal of the Royal Society Interface. 2025;**22**(231). DOI: 10.1098/rsif.2025.0263

[5] Cabanes G, Wilgenburg EV, Beekman M, Latty T. Ants build transportation networks that optimize cost and efficiency at the expense of robustness. Behavioral Ecology. 2015;**26** (1):223–231

[6] Stabentheiner A, Kovac H, Brodschneider R. Honeybee colony thermoregulation regulatory mechanisms and contribution of individuals in dependence on age, location and thermal stress. PLoS One. 2010;**5**(1):e8967

[7] Chapman RF. The Insects: Structure and Function. 5th ed. Cambridge University Press; 2012

[8] Du H, Huang R, Chen DS, Zhuang T, Huang X, Zhang H, Li Z. Regulation of soldier caste differentiation by microRNAs in Formosan subterranean termite (Coptotermes formosanus Shiraki). PeerJ. 2024;**12**:e16843. DOI: 10.7717/peerj.16843

[9] Das B, Bekker CD. Time-course RNASeq of Camponotus floridanus forager and nurse ant brains indicate links between plasticity in the biological clock and behavioral division of labor. BMC Genomics. 2022;**23**(1):57

[10] Li Q, Wang Z, Lian J, Schiøtt M, Jin L, Zhang P, Zhang Y, Nygaard S, Peng Z, Zhou Y, et al. Caste-specific RNA editomes in the leaf-cutting ant Acromyrmex echinatior. Nature Communications. 2014;**5**:4943

[11] Abreu FCPF, P.Simioes FC, P ZL. Circadian clock genes are differentially modulated during the daily cycles and chronological age in the social honeybee (Apis mellifera). Apidologie. 2018;**49** (1):71–83

[12] Mildner S, Roces F. Plasticity of daily behavioral rhythms in foragers and nurses of the ant camponotus rufipes: Influence of social context and feeding times. PLoS One. 2017;**12**(1): e0169244. DOI: 10.1371/journal.pone.0169244

[13] Boomsma JJ, Gawne R. Superorganismality and caste differentiation as points of no return: How the major evolutionary transitions

were lost in translation. Biological Reviews of the Cambridge Philosophical Society. 2018;**93**(1):28–54. DOI: 10.1111/brv.12330

[14] Siehler O, Wang S, Bloch G. Social synchronization of circadian rhythms with a focus on honeybees. Philosophical Transactions of the Royal Society B: Biological Sciences. 2021;**376**(1835):20200342. DOI: 10.1098/rstb.2020.0342

[15] Sumpter DJ. The principles of collective animal behaviour. Philosophical Transactions of the Royal Society B: Biological Sciences. 2006;**361**(1465):5–22. DOI: 10.1098/rstb.2005.1733

[16] Walter J, Jan S, Karl E, Linsenmair. Effects of light and prey availability on nocturnal, lunar and seasonal activity of tropical nightjars. Oikos. 2003;**103**(3):627–639

[17] Odgaard FB, Kjrbo PV, Poorjam AH, Hechmi K, Luciano RM, Krebs N. Automated insect detection and biomass monitoring via AI and electrical field sensor technology. Scientific Reports. 2025;**15**(1):29858

[18] Ramirez-Esquivel F, Zeil J, Narendra A. The antennal sensory array of the nocturnal bull ant Myrmecia pyriformis. Arthropod Structure & Development. 2014;**43**(6):543–558

[19] Mitaka Y, Matsuyama S, Mizumoto N, Matsuura K, Akino T. Chemical identification of an aggregation pheromone in the termite Reticulitermes speratus. Scientific Reports. 2020;**10**(1):7424. DOI: 10.1038/s41598-020-64388-4

[20] Li Y, Yin L, Guo R, Du Y, Wang B, Liu L, Li Z, Liu W, Zhang G, An S, et al. Juvenile Hormone Involved in the Defensive Behaviors of Soldiers in Termite Reticulitermes aculabialis. Insects. 2024;**15**(2). DOI: 10.3390/insects15020130

[21] Roisin Y. Caste: Termites. In: Starr CK, editor. Encyclopedia of Social Insects. Springer, Cham; 2020. p.

[22] Aurori CM, Giurgiu AI, Conlon BH, Kastally C, Dezmirean DS, Routtu J, Aurori A. Juvenile hormone pathway in honey bee larvae: A source of possible signal molecules for the reproductive behavior of Varroa destructor. Ecology and Evolution. 2021;**11**(2):1057–1068. DOI: 10.1002/ece3.7125

[23] Mizumoto N, Fuchikawa T, Matsuura K. Pairing strategy after today's failure: Unpaired termites synchronize mate search using photic cycles. Population Ecology. 2017;**59**:205–211. DOI: 10.1007/s10144-017-0584-3

[24] Zhang ZY, Ren J, Chu F, Guan JX, Yang GY, Liu YT, Zhang XY, Ge SQ, Huang QY. Biochemical, molecular, and morphological variations of flight muscles before and after dispersal flight in a eusocial termite, Reticulitermes chinensis. Insect Science. 2021;**28**(1):77–92. DOI: 10.1111/1744-7917.12763

[25] Mizumoto N, Nozaki T. The significance of social interactions in synchronized swarming flight in a termite. Biology Letters. 2024;**20**(11):20240423. DOI: 10.1098/rsbl.2024.0423

[26] Penmetcha B, Ogawa Y, Ryan LA, Hart NS, Narendra A. Ocellar spatial vision in Myrmecia ants. Journal of Experimental Biology. 2021;**224**(20). DOI: 10.1242/jeb.242948

[27] Clifton GT, Holway D, Gravish N. Vision does not impact walking performance in Argentine ants. Journal of Experimental Biology. 2020;**223**(20). DOI: 10.1242/jeb.228460

[28] Schwarz S, Mangan M, Zeil J, Webb B, Wystrach A. How Ants Use Vision When Homing Backward. Current Biology. 2017;**27**(3):401–407. DOI: 10.1016/j.cub.2016.12.019

[29] Deneubourg JL, Aron S, Goss S, Pasteels JM. The self-organizing exploratory pattern of the argentine ant. Journal of Insect Behavior. 1990;**3** (2):159

[30] Czaczkes TJG, Christoph R, Francis. Trail pheromones: An integrative view of their role in colony organization. Journal Contribution. 2015;**60**:581–599. https://hdl.handle.net /10779/uos.23414927.v1

[31] Qin W, Lin S, Chen X, Chen J, Wang L, Xiong H, Xie Q, Sun Z, Wen X, Wang C. Food transport of red imported fire ants (Hymenoptera: Formicidae) on vertical surfaces. Scientific Reports. 2019;**9**(1):3283. DOI: 10.1038/s41598-019-39756-4 From NLM

[32] Graham P, Philippides A. Vision for navigation: What can we learn from ants? Arthropod Structure & Development. 2017;**46**(5):718–722. DOI: 10.1016/j.asd.2017.07.001

[33] Mersch DP, Crespi A, Keller L. Tracking individuals shows spatial fidelity is a key regulator of ant social organization. Science. 2013;**340** (6136):1090–1093. DOI: 10.1126/science.1234316

[34] Robinson GE. Regulation of division of labor in insect societies. Annual Review of Entomology. 1992;**37**:637–665. DOI: 10.1146/annurev.en.37.010192.003225

[35] Kamhi JF, Barron AB, Narendra A. Vertical lobes of the mushroom bodies are essential for view-based navigation in australian myrmecia ants. Current Biology. 2020;**30**(17):3432–3437.e3433. DOI: 10.1016/j.cub.2020.06.030

[36] Spivak M. The wisdom of the hive—the social physiology of honey bee colonies. Annals of the Entomological Society of America. 1996;**89**:907–908

[37] Meng Q, Huang R, Yang S, Jiang W, Tian Y, Dong K. An overview of the adverse impacts of old combs on honeybee colonies and recommended beekeeping management strategies. Insects. 2025;**16**(4):351. DOI: 10.3390/insects16040351

[38] Fahrbach SE, Robinson GE. Juvenile hormone, behavioral maturation, and brain structure in the honey bee. Developmental Neuroscience. 1996;**18**(1-2):102–114. DOI: 10.1159/000111474

[39] Hasenjager MJ, Hoppitt W, Cunningham-Eurich I, Franks VR, Leadbeater E. Coupled information networks drive honeybee (Apis mellifera) collective foraging. Journal of Animal Ecology. 2024;**93**(1):71–82. DOI: 10.1111/1365-2656.14029

[40] Simpson SRB. The dance language and orientation of Bees[J]. Journal of Animal Ecology. 1969;**5**(2):460–461. DOI: 10.2307/2785

[41] Meece M, Rathore S, Buschbeck EK. Stark trade-offs and elegant solutions in arthropod visual systems. Journal of Experimental Biology. 2021;**224**(4). DOI: 10.1242/jeb.215541

[42] Raine NE, Ings TC, Dornhaus A, Saleh N, Chittka LA. . In Advances in the Study of Behavior. Vol. **36**. 2006. p. 305–354. DOI: 10.1016/S0065-3454(06) 36007-X

[43] Slatosky A, Roulston T. Bees plan camping trips: Evidence that bumble bee workers intend to spend the night away from the colony. F1000Research. 2019;**8**. DOI: 10.7490/F1000RESEARCH.1117468.1

[44] Vesala R, Harjuntausta A, Hakkarainen A, Rönnholm P, Pellikka P, Rikkinen J. Termite mound architecture regulates nest temperature and correlates with species identities of symbiotic fungi. PeerJ. 2019;**6**:e6237. DOI: 10.7717/peerj.6237

[45] Wcislo WT, Arneson L, Roesch K, Gonzalez V, Fernndez H. The evolution of nocturnal behaviour in sweat bees, Megalopta genalis and M. ecuadoria (Hymenoptera: Halictidae): An escape from competitors and enemies? Biological Journal of the Linnean Society. 2015;**83**(3):377–387

[46] Knight K. Mini-ants manage vision miniaturisation. Journal of Experimental Biology. 2019;**222**(12): jeb207902. DOI: 10.1242/jeb.207902

[47] Glaser SM, Feitosa RM, Koch A, Go N, Nascimento FSD, Grüter C. Tandem communication improves ant foraging success in a highly competitive tropical habitat. Insectes Sociaux. 2021;**68**:161–172

[48] Klein BA, Klein A, Wray MK, Mueller UG, Seeley TD Sleep deprivation impairs precision of waggle dance signaling in honey bees. In: Proceedings of the National Academy of Sciences of the United States of America; 2010, (**52**), 22705–22709

[49] Stearns SC. The Evolution Of Life Histories. Taylor & Francis Group; 1992. DOI: 10.1093/oso/9780198577416.001.0001

[50] Evans TA, Forschler BT, Grace JK Biology of Invasive Termites: A Worldwide Review. Annual Review Entomology. In Vol. **58**. 2013. p. 455–474. DOI: 10.1146/annurev-ento-120811-153554

[51] Bloch G, Bar-Shai N, Cytter Y, Green R. Time is honey: Circadian clocks of bees and flowers and how their interactions may influence ecological communities. Philosophical Transactions of the Royal Society B: Biological Sciences. 2017;**372**(1734). DOI: 10.1098/rstb.2016.0256

[52] Sanders D, Frago E, Kehoe R, Patterson C, Gaston KJ. A meta-analysis of biological impacts of artificial light at night. Nature Ecology & Evolution. 2021;**5**(1):74–81. DOI: 10.1038/s41559-020-01322-x

[53] Owens ACS, Lewis SM. The impact of artificial light at night on nocturnal insects: A review and synthesis. Ecology and Evolution. 2018;**8**(22):11337–11358. DOI: 10.1002/ece3.4557

[54] Gu G, Wang Z, Li J, Dong S, Nieh JC, Tan K. Light-induced disruptions elevate norepinephrine and impair cognition in honey bees. Entomologia Generalis. 2025;**45**(2):527–536

[55] Middendorp A, Poey B, Mawhirt C, Walker D, Mochizuki Y, Wong MKL. Ant diversity and foraging across the diel cycle is unaltered by experimental exposure to artificial light at night. Biological Conservation.

2025;**311**:111439. DOI: 10.1016/j.biocon.2025.111439

[56] Pailler L, Desvignes S, Ruhland F, Pineirua M, Lucas C. Vibratory behaviour produces different vibrations patterns in presence of reproductives in a subterranean termite species. Scientific Reports. 2021;**11**(1):9902. DOI: 10.1038/s41598-021-88292-7

[57] Byrne B, de Kort SR, Pedley SM. Leafcutter ants adjust foraging behaviours when exposed to noise disturbance. PLoS One. 2022;**17**(6): e0269517. DOI: 10.1371/journal.pone.0269517

[58] Stemkovski M, Pearse WD, Griffin SR, Pardee GL, Gibbs J, Griswold T, Neff JL, Oram R, Rightmyer MG, Sheffield CS, et al. Bee phenology is predicted by climatic variation and functional traits. Ecology Letters. 2020;**23**(11):1589–1598. DOI: 10.1111/ele.13583

[59] Gotthard K, Berger D, Rohner P. Life-History evolution of insects in response to climate variation: seasonal timing versus thermal physiology. Annual Review of Entomology. 2026;**71** (1):129–148. DOI: 10.1146/annurev-ento-121423-013506

[60] Ahmed B. Potential Impact of Climate Change on Termite Distribution in Africa. British Journal of Environment and Climate Change. 2011;**1**(4):172–189

[61] Parr CL, Bishop TR. The response of ants to climate change. Global Change Biology. 2022;**28**(10): 3188–3205. DOI: 10.1111/gcb.16140

[62] Somanathan H, Krishna S, Jos E, Gowda V, Kelber A, Borges R. Nocturnal bees feed on diurnal leftovers and pay the price of day – night lifestyle transition. Frontiers in Ecology and Evolution. 2020;**8**:566964

[63] Morrow A, Smale L, Meek PD, Lundrigan B. Trade-Offs in the sensory brain between diurnal and nocturnal rodents. Brain, Behavior and Evolution. 2024;**99**(3):123–143. DOI: 10.1159/000538090

[64] Yager DD. Predator detection and evasion by flying insects. Current Opinion in Neurobiology. 2012;**22** (2):201–207. DOI: 10.1016/j.conb.2011.12.011

[65] Falchi F, Cinzano P, Duriscoe D, Kyba CC, Elvidge CD, Baugh K, Portnov BA, Rybnikova NA, Furgoni R. The new world atlas of artificial night sky brightness. Science Advances. 2016;**2**(6):e1600377. DOI: 10.1126/sciadv.1600377

Chapter 3

Physiological and Behavioral Adaptations of Scorpions to Nocturnal Life in Arid Environments

Manel Khammassi

Abstract

Scorpions represent one of the most ancient lineages of terrestrial arthropods, exhibiting remarkable physiological and behavioral adaptations that allow them to thrive as nocturnal predators in some of the planet's harshest environments. Their nocturnality is a key evolutionary strategy to minimize water loss, avoid thermal stress, and optimize prey capture. This chapter explores how scorpions combine inherited traits and adaptive plasticity to cope with fluctuating nocturnal conditions. Particular emphasis is given to their sensory specializations, including trichobothria, pectines, and cuticular fluorescence, which facilitate navigation and prey detection in low-light conditions. The chapter also highlights physiological mechanisms regulating metabolism, desiccation resistance, and venom efficiency under varying thermal regimes. By integrating recent advances in scorpion behavioral ecology, molecular physiology, and evolutionary biology, this contribution provides insights into how nocturnality and environmental plasticity interact to shape the ecological success and resilience of scorpions in arid and semi-arid ecosystems. These findings also underscore the relevance of scorpions as bioindicators of environmental change in desert habitats.

Keywords: scorpions, adaptations, nocturnal life, arid environments, behavior

1. Introduction

Nocturnality represents a major adaptive strategy among terrestrial organisms inhabiting arid and semi-arid environments. In deserts and steppes, extreme daytime temperatures, intense solar radiation, and low humidity impose severe physiological constraints. Shifting activity to nighttime allows organisms to reduce water loss, avoid thermal stress, and exploit prey resources with reduced competition and predation pressure [1–3].

Among arthropods, nocturnal activity is particularly widespread and has driven the evolution of specialized sensory, physiological, and behavioral traits enabling efficient functioning under low light and cooler conditions [4, 5].

IntechOpen

Scorpions represent one of the most ancient lineages of terrestrial arthropods, with a fossil record dating back over 430 million years (Silurian) [6]. Such, the fossil scorpion, *Parioscorpio venator* Wendruff *et al.*, 2020, is the oldest scorpion known to date (437.5–436.5 Ma), found in Wisconsin, USA [7].

Their long evolutionary history, combined with their strict nocturnal habits and predominance in arid ecosystems, makes them ideal model organisms for studying nocturnal adaptation. Unlike many nocturnal insects, scorpions are obligate nocturnal predators throughout their life cycle. Their survival depends on the integration of sensory acuity, physiological tolerance, and behavioral flexibility, allowing them to persist in some of the harshest terrestrial environments on Earth [4].

Scorpions occupy a wide range of arid habitats, including deserts, steppes, rocky plateaus, and coastal dunes. Genera such as *Androctonus*, *Buthus*, *Scorpio*, *Leiurus* and *Buthacus* dominate arid faunas in North Africa, the Middle East, and Central Asia [8, 9]. Their success is largely attributed to their nocturnal lifestyle, low metabolic rates, high desiccation resistance, and burrowing behavior, which collectively buffer them against climatic extremes [10–12].

This chapter aims to synthesize current knowledge on the physiological, morphological, sensory, and behavioral adaptations that enable scorpions to thrive as nocturnal predators in arid environments. Emphasis is placed on concrete examples, with an evolutionary perspective integrating phylogenetic and ecological evidence.

2. Evolutionary background and phylogenetic insights

2.1 Origin and diversification of scorpions

Scorpions originated during the Paleozoic, with early forms likely occupying marginal aquatic or amphibious environments [13, 14]. The transition to fully terrestrial life involved major evolutionary innovations, including book lungs for aerial respiration, a waxy epicuticle to reduce water loss, and modified appendages for prey capture and burrowing [15–20]. Subsequent diversification during the Mesozoic and Cenozoic coincided with the expansion of arid and semi-arid habitats, particularly in Gondwana and later Afro-Arabian regions [21–24].

2.2 Transition from aquatic to terrestrial life

The colonization of land required scorpions to overcome desiccation, gravity, and thermal variability. Comparative studies suggest that scorpions retained conservative physiological traits, such as low metabolic rates and discontinuous gas exchange, which later facilitated adaptation to arid terrestrial environments [25–28]. Burrowing behavior, already present in early terrestrial forms, became a key strategy for avoiding daytime heat and maintaining stable microclimates underground [29].

2.3 Evolutionary history drivers of nocturnality and arid adaptation

Nocturnality in scorpions is considered an ancestral trait reinforced by selection in hot environments. Phylogenetic analyses indicate that strictly nocturnal activity predates the radiation of many desert-adapted lineages, suggesting that nocturnality provided a pre-adaptation to aridity [21, 22, 30].

Increased aridification during the Miocene further promoted ecological specialization, favoring species capable of exploiting nocturnal niches while minimizing water loss and thermal exposure [31–33].

In the genus *Scorpio*, lineage divergence correlates with historical aridification events and habitat fragmentation, suggesting long-term stability of nocturnal and burrowing strategies across evolutionary timescales [33].

3. Morphological and sensory adaptations

3.1 Exoskeleton and cuticular fluorescence

The scorpion exoskeleton plays a critical role in protection, water retention, and sensory perception. Its thick, wax-coated cuticle reduces evaporative water loss, a crucial adaptation in arid environments [12, 34].

A distinctive feature of scorpions is their cuticular fluorescence under ultraviolet light, caused by β-carboline and coumarin compounds embedded in the cuticle [35–37]. While the adaptive significance remains debated, proposed functions include UV detection, camouflage under moonlight, and intra-specific signaling [34].

3.2 Sensory organs

3.2.1 Trichobothria and mechanoreception

Scorpions rely strongly on mechanoreception to detect prey and threats in darkness. Trichobothria, specialized sensory hairs located on the pedipalps, are highly sensitive to air movements and vibrations, enabling accurate prey localization. Their number and arrangement vary among taxa and are taxonomically informative [38]. In sand-dwelling species such as *Paruroctonus mesaensis* and *Heterometrus petersii*, prey detection also involves basitarsal compound slit sensilla (BCSS) that perceive substrate-borne surface vibrations. These sensilla allow scorpions to determine prey direction and distance, with precise localization requiring integration of signals from multiple legs. In addition, tarsal sensory hairs respond to high-frequency compressional waves, complementing vibration-based prey detection in sandy substrates [39–41].

3.2.2 Pectines: Chemosensory and mechanosensory functions

Pectines are unique ventral appendages exclusive to scorpions and serve as chemoreceptive and a mechanoreceptive organs. They are involved in substrate exploration, such humid environments detection, mate recognition, and chemical cue detection [42–45].

In nocturnal desert species such as *Leiurus quinquestriatus*, pectines play a key role in assessing soil texture and chemical composition, influencing microhabitat selection and burrow placement [46–47]. Experimental studies have shown a sand preference in scorpions, and an altered preference when the pectines were covered with wax [48].

3.2.3 Median and lateral eyes

Although scorpion vision is relatively poor compared to diurnal arthropods, their median and lateral eyes are adapted to low-light conditions. Schliwa and Fleissner (1980) showed that scorpion median and lateral eyes have distinct functions: median eyes provide higher visual acuity and spatial discrimination, while lateral eyes are extremely sensitive to low light and primarily act as brightness detectors involved in circadian rhythm synchronization [49]. Behavioral experiments indicate that scorpions use visual cues primarily for orientation and predator avoidance rather than prey capture [50].

3.2.4 Locomotion and burrowing behavior

Efficient locomotion and burrowing are central to nocturnal survival. Scorpions exhibit low-energy, stop-and-go movement patterns that reduce metabolic costs and limit exposure to predators [51, 52].

Burrowing behavior provides thermal buffering, humidity retention, and protection during daylight hours. In *Urodacus* and *Scorpio* species, burrows can exceed 1 m in depth and maintain stable temperature and humidity conditions even when surface temperatures + 45 °C [53, 54]. Similar strategies are observed in other desert scorpions inhabiting sandy or rocky substrates, highlighting the tight link between morphology, behavior, and edaphic conditions [29, 55].

4. Physiological mechanisms of adaptation

4.1 Water conservation mechanisms

Water availability represents the principal limiting factor in arid environments, and scorpions have evolved highly efficient physiological mechanisms to minimize water loss. Central to this adaptation is the exceptionally low permeability of the integument, as evaporative water loss (EWL) in scorpions occurs predominantly through the cuticle. The epicuticle is coated with a thick lipid layer rich in cuticular hydrocarbons (HCs), which forms the primary barrier to water vapor diffusion and confers some of the lowest recorded water-loss rates (WLRs) among terrestrial arthropods [56–58].

Cuticular waterproofing in scorpions is not static but exhibits notable plasticity in response to desiccation stress. Experimental studies on the desert scorpion *Buthus occitanus* have shown that acute exposure to dry conditions leads to an increase in the total quantity of cuticular HCs, accompanied by a shift in HC composition toward longer carbon chains and a higher proportion of branched alkanes. These changes closely mirror seasonal variations observed in field-collected scorpions and are associated with enhanced resistance to water loss [15, 59]. The absence of unsaturated hydrocarbons—common in many insects—further contributes to the high melting points and low permeability of the scorpion cuticle [60].

Moreover, scorpions possess unusually high concentrations of cuticular sterols, particularly cholesterol, which may stabilize the hydrocarbon matrix and reduce lipid fluidity, thereby reinforcing the barrier function of the epicuticle. Together, these structural and compositional properties allow scorpions to maintain water balance during prolonged periods of drought, despite limited access to free water and infrequent feeding opportunities [58].

In addition to their highly impermeable cuticle, scorpions possess efficient excretory and respiratory adaptations that further reduce water loss. Nitrogenous wastes are eliminated primarily in the form of insoluble guanine and uric acid, rather than urea or ammonia, thereby minimizing water loss associated with excretion [2, 12]. This mode of excretion is particularly advantageous for desert-dwelling and nocturnal species, such as members of the family Buthidae, which may experience prolonged periods without access to free water.

Respiratory water loss (RWL), although generally a minor component of total evaporative water loss in scorpions, is also tightly regulated. Comparative studies have shown that xeric species —those adapted to dry, arid habitats— such as *Leiurus quinquestriatus*, exhibit significantly lower RWL and reduced H_2O/CO_2 emission ratios than mesic species, which inhabit relatively moist environments, reflecting enhanced gas exchange efficiency and tighter control of book lung ventilation [61–63]. At rest, RWL may account for only ~10% of total water loss in xeric scorpions, whereas higher proportions are observed in mesic species or during periods of elevated activity [62, 63].

Physiologically, reduced RWL in xeric scorpions may be achieved through elevated hemolymph PCO_2, which increases CO_2 washout efficiency while limiting water loss during gas exchange [61]. This mechanism appears to be further tightened during prolonged desiccation. However, RWL is strongly influenced by activity patterns. For example, the mesic scorpion *Scorpio maurus fuscus* shows low resting RWL but high activity levels, leading to respiratory losses of up to 30% of daily EWL —a much higher contribution than in other species. This highlights the importance of behavioral regulation (reduced activity, sheltering) as a complementary water-conservation strategy [62, 63].

Beyond water loss reduction, scorpions differ in how they store and mobilize body water. During desiccation, water is lost primarily from the hepatopancreas, while hemolymph volume and osmolarity are tightly regulated, especially in buthids. Xeric buthids such as *L. quinquestriatus* store large amounts of water in the hepatopancreas and mobilize these reserves efficiently, maintaining osmotic stability even under severe dehydration. In contrast, scorpionids exhibit higher initial WLRs, rapid depletion of hepatopancreatic water, poorer osmoregulatory capacity, and higher mortality under extreme desiccation [11, 64].

The hepatopancreas thus functions as a central water reservoir, storing both bulk water and metabolically bound water (e.g. glycogen-associated). In species with high WLRs, carbohydrate catabolism may release bound water but leads to rapid depletion of reserves and limited tolerance to prolonged drought [65, 66].

4.2 Metabolic plasticity and dormancy

Metabolic plasticity plays a central role in the ability of scorpions to survive prolonged periods of thermal and hydric stress characteristic of arid and semi-arid environments. Several physiological studies demonstrate that scorpions combine intrinsically low metabolic rates with a strong capacity to modulate energy expenditure in response to environmental conditions, thereby reducing both energetic and water demands.

Early comparative work by Robertson et al. [11] showed that African scorpions maintain tight osmoregulatory control across a range of temperatures, with only modest increases in oxygen consumption as temperature rises. Importantly, metabolic rate and water loss were not tightly coupled, indicating that scorpions can elevate metabolic activity when necessary without proportionally increasing evaporative water loss. This decoupling supports short-term physiological adjustments that conserve body water under fluctuating desert temperatures.

Direct measurements of metabolic rate confirmed that scorpions rank among the lowest metabolic arthropods known. Using flow-through respirometry, Lighton et al. [25] demonstrated exceptionally low resting metabolic rates in several scorpion species, even when compared with other sit-and-wait predators. Such metabolic suppression reduces food and water requirements and allows scorpions to persist at relatively high population densities despite limited prey availability. The authors further suggested that low metabolism promotes tolerance to extended fasting and may underlie the frequent occurrence of cannibalism as a population-regulating mechanism in resource-poor habitats.

Temperature-dependent metabolic flexibility was further quantified by Withers and Smith [67] in the Australian scorpion *Urodacus armatus*. Their results showed a shallow thermal sensitivity of metabolic rate (low Q_{10}), indicating that oxygen consumption increases only moderately with temperature. Although evaporative water loss rose with warming, the increase remained low compared with that of many insects, suggesting that metabolic plasticity is coupled with effective control of water loss. At low temperatures, scorpions exhibited profound metabolic depression, consistent with a state of functional dormancy that minimizes energy use during unfavorable periods.

Together, these studies indicate that scorpions rely on a combination of constitutively low metabolism, temperature-insensitive metabolic scaling, and reversible metabolic depression to endure extreme environmental stress. Rather than true diapause, scorpions enter a flexible, dormancy-like physiological state during prolonged inactivity, allowing them to survive extended periods of food and water scarcity while maintaining readiness for rapid activation when conditions improve.

4.3 Thermoregulation strategies

Scorpions are ectothermic arthropods that rely primarily on behavioral thermoregulation rather than physiological control to cope with extreme thermal conditions. Species from xeric habitats tolerate and prefer higher temperatures than mesic species, but still actively avoid thermal extremes that would increase metabolic demands and water loss [68]. Field and laboratory studies on *Centruroides vittatus* show that scorpions regulate body temperature through temporal shifts in activity and selective microhabitat use. Individuals remain inactive during the day and occupy thermally buffered refuges such as burrows, rock crevices, and woody debris, becoming active at night when surface temperatures are lower [69, 70]. This behavior simultaneously reduces heat stress and desiccation risk.

Thermoregulation is tightly linked to hydration state and survival. Elevated temperatures significantly increase water loss rates and constrain surface activity, thereby influencing species distribution patterns [71]. Moreover, dehydration and heat stress can impair antipredator responses, revealing trade-offs between thermal regulation, water balance, and defensive behavior [72]. Collectively, these behavioral strategies enable scorpions to maintain thermal safety margins while minimizing energetic and hydric costs in arid and variable environments.

4.4 Venom adaptation and energetic efficiency

Venom production is energetically costly, and scorpions have evolved energy-efficient venom strategies adapted to nocturnal predation. Venom composition varies among species, reflecting trade-offs between toxicity, volume, and energetic cost [73, 74]. Nocturnal scorpions typically rely on rapid immobilization of prey through neurotoxic venom components, minimizing struggle time and reducing energy expenditure during prey handling [75]. In arid environments, efficient venom use is particularly advantageous, as it limits water loss and physical damage during predation.

5. Behavioral ecology of nocturnality

5.1 Circadian rhythms and activity patterns

Behavioral regulation plays a central role in complementing the physiological adaptations of scorpions to arid and thermally extreme environments. Scorpions typically restrict surface activity to nocturnal periods and remain inactive during the hottest and driest parts of the day, retreating into burrows, rock crevices, leaf litter, or woody debris that provide thermally and hygrometrically buffered microclimates [5, 51]. These refugia reduce exposure to lethal temperatures, limit evaporative water loss, and allow metabolic downregulation during daytime inactivity.

The importance of behavioral control of activity is particularly evident in desert-adapted species, where surface activity is tightly constrained by hydric costs. Experimental and field studies have shown that increased activity substantially elevates respiratory and cuticular water loss, which can account for a significant proportion of daily evaporative losses under warm conditions [67, 71]. For example, in *Scorpio maurus fuscus* and *Buthotus judaicus*, elevated temperatures rapidly increase water loss rates, effectively limiting diurnal activity and reinforcing nocturnality as a water-conserving strategy [66, 71].

Scorpion activity is further regulated by endogenous circadian rhythms, with peak emergence typically occurring shortly after sunset and declining toward dawn [5]. These rhythms are flexible and modulated by exogenous cues such as temperature, humidity, wind, and lunar illumination. Studies on *Paruroctonus utahensis* and *Centruroides vittatus* demonstrate that surface activity is reduced during unfavorable conditions, including cold nights, high winds, and periods of low humidity [69, 70, 76]. Seasonal variation is also pronounced: in Mediterranean and Saharan species, activity peaks during warm, humid nights in spring and autumn, whereas extreme summer heat and winter cold lead to prolonged inactivity or retreat to deeper refuges [4, 77].

Predation risk further shapes nocturnal activity patterns. Moonlight intensity has been shown to significantly influence foraging behavior, with scorpions reducing surface activity during bright moonlit nights to avoid visually oriented predators such as birds and mammals [78]. This state-dependent foraging reflects a trade-off between energy acquisition and survival, reinforcing the adaptive value of nocturnality in open habitats.

5.2 Predatory Strategies, Sensory-Guided Foraging and Cannibalism

Scorpions are predominantly sit-and-wait predators, a strategy that minimizes energetic expenditure and complements their exceptionally low metabolic rates [34]. Prey detection relies on highly specialized mechanosensory systems, including trichobothria on the pedipalps and slit sensilla sensitive to substrate-borne vibrations, enabling precise localization of prey in complete darkness [39]. Once detected, prey is subdued rapidly using coordinated strikes of the pedipalps and metasoma, often within seconds. Nocturnal activity enhances prey encounter rates, as many arthropod prey species are themselves most active at night [51].

Nocturnality also reduces exposure to diurnal predators; however, scorpions remain vulnerable to nocturnal predators such as small mammals, owls, and other arthropods. Defensive strategies include cryptic coloration, prolonged immobility, rapid retreat to refuges, and venom deployment when threatened [72, 79]. Dehydration and thermal stress have been shown to impair defensive responses, underscoring the tight linkage between physiological state and behavior [72]. Cannibalism, including sexual and juvenile cannibalism, may occur under conditions of food scarcity, but its frequency is mitigated by low metabolic demands and infrequent foraging requirements [25].

5.3 Reproductive behavior and parental care

Reproductive activities in scorpions are also closely synchronized with nocturnality. Courtship and mating occur almost exclusively at night and involve complex tactile and chemical communication, including the characteristic "promenade à deux", during which males guide females to spermatophore deposition sites [80]. Nocturnal mating reduces thermal and hydric stress while limiting exposure to predators during prolonged courtship.

Following fertilization, females exhibit extended parental care, carrying juveniles on their dorsum during early developmental stages. This behavior enhances offspring survival in harsh environments and represents a significant energetic investment that is facilitated by prolonged daytime sheltering and burrowing behavior [5].

6. Environmental plasticity and climate resilience

Scorpions exhibit a high degree of environmental plasticity, allowing them to adjust activity patterns, burrow depth, and metabolic rates in response to fluctuating temperature and humidity [2]. This plasticity is expressed primarily through behavioral adjustments, including shifts in activity timing, flexible use of microhabitats, and changes in refuge depth, which together reduce exposure to thermal and desiccation stress [5, 51]. Species occupying heterogeneous landscapes can rapidly suppress surface activity during unfavorable conditions and resume foraging when microclimatic conditions improve [76].

Physiological traits reinforce this flexibility. Scorpions possess extremely low basal metabolic rates and the ability to downregulate oxygen consumption under stress, enabling survival during prolonged drought or prey scarcity [11, 25]. However, metabolic and hydric tolerance is not unlimited. Experimental studies show that water loss increases sharply above critical temperature thresholds, constraining activity and ultimately influencing habitat occupancy and geographic distribution [67, 71].

Climate change is expected to intensify thermal extremes and aridity across many scorpion habitats. Short-term resilience may be achieved through enhanced nocturnality, altered activity rhythms, and microhabitat switching, but long-term persistence will depend on physiological tolerance limits, dispersal capacity, and the availability of suitable refuges [68, 72]. Consequently, species with broad ecological tolerance are likely to persist under changing climates, whereas habitat specialists may face heightened risk of population decline or local extinction.

7. Ecological and conservation implications

7.1 Role of scorpions in desert food webs and ecosystem functioning

Scorpions occupy a pivotal position in desert food webs, functioning both as mesopredators and prey. As generalist predators, they exert significant top-down control on populations of insects and other arthropods, including termites, orthopterans, beetles, and spiders [4]. Through this regulatory role, scorpions contribute to maintaining arthropod community structure and preventing population outbreaks, particularly in resource-poor arid systems where trophic cascades can be pronounced [81].

Beyond predation, scorpions indirectly contribute to nutrient cycling in desert soils. Their burrowing activity modifies soil structure, aeration, and water infiltration, enhancing microhabitat heterogeneity and facilitating microbial processes [79]. Burrows accumulate organic debris and prey remains, creating localized nutrient hotspots that benefit soil fauna and microorganisms. Such bioturbation effects are especially relevant in arid ecosystems, where nutrient turnover is slow and spatially heterogeneous.

Scorpions also constitute an important prey resource for nocturnal vertebrates such as owls, foxes, hedgehogs, and small mammals, thereby linking invertebrate and vertebrate trophic levels [82, 83]. Their ecological role thus extends beyond simple predator–prey interactions, contributing to ecosystem stability and energy flow.

7.2 Scorpions as bioindicators of environmental disturbance and climate variability

Due to their strong dependence on microclimatic conditions, soil properties, and burrowing habitats, scorpions are increasingly recognized as potential bioindicators of environmental change. Their sedentary lifestyle, limited dispersal capacity, and long lifespan make them particularly sensitive to habitat alteration and climatic shifts [4, 21, 22].

Changes in scorpion assemblage composition, abundance, or activity patterns can reflect alterations in soil moisture, salinity, temperature regimes, and vegetation cover. For example, declines in moisture-dependent burrowing species have been documented in areas affected by desertification and soil degradation, while opportunistic or xerophilic species may expand under increasing aridity [84]. Similarly, shifts in phenology—such as delayed or reduced nocturnal activity—may signal rising night-time temperatures linked to climate change.

At broader spatial scales, scorpion distribution patterns have been shown to track bioclimatic gradients and historical landscape stability, making them useful indicators in biogeographic and conservation assessments [85]. Integrating scorpions into long-term monitoring programs could therefore provide valuable insights into ecosystem health in arid and semi-arid regions.

7.3 Major threats: Habitat loss, climate change, and overcollection

Despite their resilience, scorpions face growing threats from both anthropogenic and environmental pressures. Habitat loss and fragmentation, driven by agricultural expansion, urbanization, and infrastructure development, directly reduce suitable burrowing habitats and disrupt population connectivity. Soil compaction, pollution, and alteration of hydrological regimes further degrade microhabitats critical for survival and reproduction.

Climate change represents a more pervasive and long-term threat. Rising temperatures, increased frequency of heatwaves, and prolonged droughts may exceed physiological tolerance limits, particularly for species adapted to narrow thermal or edaphic niches. While nocturnality and behavioral plasticity may buffer short-term impacts, sustained shifts in temperature and humidity could alter activity windows, prey availability, and reproductive success [86].

In some regions, overcollection for the pet trade, traditional medicine, or venom extraction poses an additional threat, especially for range-restricted or charismatic species. Unsustainable harvesting can rapidly deplete local populations, given the low reproductive rates and limited dispersal typical of many scorpion species [85].

7.4 Conservation priorities and future research directions

Effective conservation of scorpions requires an integrated approach that recognizes their ecological specificity and evolutionary history. Protecting soil integrity and microhabitat diversity is paramount, as burrowing substrates and microclimatic refuges are central to scorpion survival. Conservation strategies should therefore prioritize habitat quality over species-focused interventions alone.

From a research perspective, future studies should aim to:

- Quantify physiological tolerance limits to temperature and desiccation across taxa;
- Integrate behavioral ecology with population genetics and phylogeography;
- Assess the long-term impacts of climate change on nocturnal activity patterns;

- Develop standardized protocols for using scorpions as bioindicators in arid ecosystems.
- Such integrative approaches will be essential for predicting species' responses to environmental change and for designing effective conservation measures.

8. Conclusion

Scorpions exemplify how the interplay between physiology, behavior, and evolution can generate successful adaptive strategies in some of the harshest environments on Earth. Nocturnality emerges as a central organizing trait, reducing thermal and hydric stress while shaping sensory systems, metabolic strategies, and ecological interactions. These adaptations are not isolated responses but part of a tightly integrated adaptive syndrome refined over hundreds of millions of years.

The balance between inherited traits—such as cuticular impermeability, venom efficiency, and sensory specialization—and phenotypic plasticity—including behavioral thermoregulation, activity timing, and burrow use—underpins the remarkable resilience of scorpions in arid ecosystems. However, this balance is not unlimited; accelerating environmental change may push some species beyond their adaptive capacity.

More broadly, understanding nocturnal adaptation in scorpions provides valuable insights into how terrestrial organisms cope with extreme environments. As global aridification intensifies, scorpions offer a powerful model for exploring the limits of adaptation, the role of historical contingency, and the mechanisms by which life persists under environmental constraint.

Acknowledgments

The author would like to thank the publisher and the academic editor for the opportunity to take part in this work.

Conflict of Interest

The author declares no conflict of interest.

Author details

Manel Khammassi
Laboratory of Sylvo-Pastoral Resources, Sylvo-Pastoral Institute of Tabarka, University of Jendouba, Tabarka, Tunisia

*Address all correspondence to: manel.khammassi@fst.utm.tn

IntechOpen

© 2026 The Author(s). Licensee IntechOpen. This chapter is distributed under the terms of the Creative Commons Attribution License (http://creativecommons.org/licenses/by/4.0/), which permits unrestricted use, distribution, and reproduction in any medium, provided the original work is properly cited.

References

[1] Charles-Dominique P. Nocturnality and Diurnality: An ecological interpretation of these two modes of life by an analysis of the higher vertebrate fauna in tropical forest ecosystems. In: Luckett WP, Szalay FS, editors. Phylogeny of the Primates. Boston, MA: Springer; 1975. p. 69–88. DOI: 10.1007/978-1-4684-2166-8_4

[2] Cloudsley-Thompson JL. Ecophysiology of Desert Arthropods and Reptiles. Springer Berlin, Heidelberg: Springer Science & Business Media; 2012. p. 203. DOI: 10.1007/978-3-642-75337-4

[3] Daily GC, Ehrlich PR. Nocturnality and species survival. Proceedings of the National Academy of Sciences of the United States of America. 1996;**93**(21):11709–11712

[4] Polis G. Ecology. In: Polis GA, editor. The Biology of Scorpions. Stanford, CA: Stanford University Press; 1990. p.123–144

[5] Warburg MR, Polis GA. Behavioral responses, rhythms, and activity patterns. In: Polis GA, editor. The Biology of Scorpions. Stanford, CA: Stanford University Press; 1990. p. 224–246

[6] Dunlop JA, Selden PA. Scorpion fragments from the Silurian of Powys, Wales. Arachnology. 2013;**16**(1):27–32. DOI: 10.13156/arac.2013.16.1.27

[7] Wendruff AJ, Babcock LE, Wirkner CS, Kluessendorf J, G MD. A Silurian ancestral scorpion with fossilised internal anatomy illustrating a pathway to arachnid terrestrialisation. Scientific Reports. 2020;**10**(1):1–6

[8] Lourenço WR. The scorpion families and their geographical distribution. Journal of Venomous Animals and Toxins. 2001;7(1):03–23

[9] Lourenço WR. The evolution and distribution of noxious species of scorpions (Arachnida: Scorpiones). Journal of Venomous Animals and Toxins Including Tropical Diseases. 2018;**24**(1). DOI: 10.1186/s40409-017-0138-3

[10] Polis GA. The effect of cannibalism on the demography and activity of a natural population of desert scorpions. Behavioral Ecology and Sociobiology. 1980;7(1):25–35. DOI: 10.1007/BF00302515

[11] Robertson HG, Nicolson SW, Louw GN. Osmoregulation and temperature effects on water loss and oxygen consumption in two species of African scorpion. Comparative Biochemistry and Physiology Part A: Physiology. 1982;**71**(4):605–609

[12] Hadley NF. The Water Relations of Terrestrial Arthropods. San Diego: Academic Press; 1994

[13] Jeram AJ. Carboniferous Orthosterni and their relationship to living scorpions. Palaeontology. 1994;**37**:513–550

[14] Jeram AJ Phylogeny, classification and evolution of Silurian and Devonian scorpions. In: Selden PA, editor. Proceedings of the 17th European Colloquium of Arachnology, Edinburgh; 1997

[15] Toolson EC, Hadley NF. Seasonal effects on cuticular permeability and

epicuticular lipid composition in Centruroides sculpturatus ewing 1928 (Scorpiones: Buthidae). Journal of Comparative Physiology B: Biochemical, Systemic, and Environmental Physiology. 1979;**129**:319–325. DOI: 10.1007/BF00686988

[16] Selden PA, Jeram AJ. Palaeophysiology of terrestrialisation in the Chelicerata. Earth and Environmental Science Transactions of the Royal Society of Edinburgh. 1989;**80** (3-4):303–310. DOI: 10.1017/S0263593300028741

[17] Dunlop JA, Legg DA, Selden PA, Fet V, Schneider JW, Rößler R. Permian scorpions from the Petrified Forest of Chemnitz, Germany. BMC Evolutionary Biology. 2016;**16**(1):1–16. DOI: 10.1186/s12862-016-0634-z

[18] Howard RJ, Edgecombe GD, Legg DA, Pisani D, Lozano-Fernandez J. Exploring the evolution and terrestrialization of scorpions (Arachnida: Scorpiones) with rocks and clocks. Organisms Diversity and Evolution. 2019;**19**(1):71–86. DOI: 10.1007/s13127019-00390-7

[19] Haug C, Wagner P, Haug JT. The evolutionary history of body organisation in the lineage towards modern scorpions. Bulletin of Geosciences. 2019;**94**(4):389–408. DOI: 10.3140/bull.geosci.1750

[20] Haug C. The evolution of feeding within Euchelicerata: Data from the fossil groups Eurypterida and Trigonotarbida illustrate possible evolutionary pathways. PeerJ. 2020;**8**: e9696. DOI: 10.7717/peerj.9696

[21] Prendini L. Chapter: Substratum specialization and speciation in Southern African scorpions: The Effect Hypothesis revisited. In: Fet V, Selden PA, editors. Scorpions 2001. In Memoriam Gary A. Polis. Burnham Beeches, Bucks: British Arachnological Society; 2001. p. 113–138

[22] Prendini L. Systematics, Evolution and Biogeography of the Southern African Burrowing Scorpions, Opistophthalmus CL Koch (Scorpiones, Scorpionidae). Ph. D. thesis, University of Cape Town; 2001

[23] Lourenço WR. Panbiogéographie, les distributions disjointes and le concept de famille relictuelle chez les scorpions. Biogeographica (Paris). 1998;**74**(3):133–144

[24] Sharma PP, Baker CM, Cosgrove JG, Johnson JE, Oberski JT, Raven RJ, Harvey MS, Boyer SL, Giribet G. A revised dated phylogeny of scorpions: Phylogenomic support for ancient divergence of the temperate Gondwanan family Bothriuridae. Molecular Phylogenetics and Evolution. 2018;**122**:37–45. DOI: 10.1016/j.ympev.2018.01.003

[25] Lighton JR, Brownell PH, Joos B, Turner RJ. Low metabolic rate in scorpions: Implications for population biomass and cannibalism. Journal of Experimental Biology. 2001;**204** (3):607–613

[26] Dunlop JA, Webster M. Fossil evidence, terrestrialization and arachnid phylogeny. The Journal of Arachnology. 1999;**27**(1):86–93

[27] Lozano-Fernandez J, Tanner AR, Giacomelli M, Carton R, Vinther J, Edgecombe GD, Pisani D. Increasing species sampling in chelicerate genomic-scale datasets provides support for monophyly of Acari and Arachnida.

Nature communications. 2019;**10**(1):1–8 DOI: 10.1038/s41467-019-10244-7

[28] Lozano-Fernandez, J, Tanner AR, Puttick MN, Vinther J, Edgecombe GD, Pisani D. A Cambrian–Ordovician terrestrialization of arachnids. Frontiers in Genetics. 2020;**11**:182. DOI: 10.3389/fgene.2020.00182

[29] Adams AM, Marais E, Turner JS, Prendini L, Pinshow B. Similar burrow architecture of three arid-zone scorpion species implies similar ecological function. The Science of Nature. 2016;**103**(7):1–11. DOI: 10.1007/s00114-016-1374-z

[30] Polis GA, Farley RD. Behavior and ecology of mating in the cannibalistic scorpion, Paruroctonus mesaensis Stahnke (Scorpionida: Vaejovidae). Journal of Arachnology. 1979;7(1):33–46

[31] Sousa P, Harris DJ, Froufe E, Van Der Meijden A. Phylogeographic patterns of Buthus scorpions (Scorpiones: Buthidae) in the Maghreb and South-Western Europe based on CO1 mtDNA sequences. Journal of Zoology. 2012;**288**(1):66–75

[32] Shi CM, Ji YJ, Liu L, Wang L, Zhang DX. Impact of climate changes from Middle Miocene onwards on evolutionary diversification in Eurasia: insights from the mesobuthid scorpions. Molecular Ecology. 2013;**22**(6):1700–1716

[33] Khammassi M, Nouira S, Badry A, Sadine SE, Harris DJ. Phylogeography and evolutionary history of the burrowing scorpion genus Scorpio Linnaeus, 1758 (Scorpiones: Scorpionidae) in the Mediterranean Basin. Organisms Diversity & Evolution. 2014;**24**(3)335–352

[34] Gaffin DD, Bumm LA, Taylor MS, Popokina NV, Mann S. Scorpion fluorescence and reaction to light. Animal Behaviour.2012;**83**(2):429–436

[35] Anglade F, Ricordel I, Goyffon M. Données spectroscopiques sur la fluorescence de la cuticule de scorpion. Bulletin de la Société Européenne d'Arachnologie. 1990;**1**:5–9

[36] Frostl LM, Butlerl DR, O'Dell B, Fet V. LM, Butlerl DR, O'Dell B, Fet V, A coumarin as a fluorescent compound in scorpion cuticle. In: Fet V, Selden PA, editors. Scorpions 2001. In Memoriam Gary A. Polis. Bunham Beeches, Bucks: British Arachnological Society; 2001. p. 365–368

[37] Yoshimoto Y, Tanaka M, Miyashita M, Abdel-Wahab M, Megaly AM, Nakagawa Y, Miyagawa H. A Fluorescent Compound from the Exuviae of the Scorpion, Liocheles australasiae. Journal of Natural Products. 2020;**83**(2):542–546. DOI: 10.1021/acs.jnatprod.9b00972

[38] Vachon M. Etude des caractères utilisés pour classer les familles and les genres de Scorpions (Arachnides). 1. La trichobothriotaxie en arachnologie. Sigles trichobothriaux and types de trichobothriotaxie chez les Scorpions. Bulletin du Muséum National d'Histoire Naturelle, Paris. 1974;**140**:857–958

[39] Brownell PH. Compressional and surface waves in sand: used by desert scorpions to locate prey. Science. 1977;**197**:479–482. DOI: 10.1126/science.197.4302.479

[40] Brownell P, Farley RD. Detection of vibrations in sand by tarsal sense organs of the nocturnal scorpion,

Paruroctonus mesaensis. Journal of comparative physiology. 1979;**131** (1):23–30. DOI: 10.1007/BF00613080

[41] Wang K, Zhang J, Liu L, Chen D, Song H, Wang Y, Niu S, Han Z, Ren L. Vibrational receptor of scorpion (Heterometrus petersii): The basitarsal compound slit sensilla. Journal of Bionic Engineering. 2019;**16**(1):76–87. DOI: 10 .1007/s42235-019-0008-5

[42] Ivanov V, Balashov YS. The structural and functional organization of the pectine in a scorpion Buthus eupeus Koch (Scorpiones, Buthidae) studied by electron microscopy. The fauna and ecology of Arachnida. 1979;**85**:73–87

[43] Foelix RF, Muller-Vorholt GRF, Muller-Vorholt G. Fine structure of scorpion sensory organs. II. Pecten sensilla. Bulletin-British Arachnological Society. 1983;**6**(2):68–74

[44] Drozd D. Topographic organization of the pectine neuropils in scorpions: An analysis of chemosensory afferents and the projection pattern in the central nervous system. Springer Spektrum; 2019.

[45] Drozd D, Wolf H, Stemme T. Mechanosensory pathways of scorpion pecten hair sensillae - Adjustment of body height and pecten position. Journal of Comparative Neurology. 2022;**530** (16):2918–2937. DOI: 10.1002/ cne.25384

[46] Abushama FT. On the behaviour and sensory physiology of the scorpion Leiurus quinquestriatus (H. & E.). Animal behaviour. 1964;**12**(1):140–153

[47] Carthy JD. Fine structure and function of the sensory pegs on the scorpion pectine. Experientia. 1966;**22**: 89. DOI: 10.1007/BF01900166

[48] Root TM. Neurobiology. In: Polis GA, editor. The Biology of Scorpions. Stanford University Press. Stanford, CA; 1990. 2018;**24**(1):167–199. DOI: 10.1186/s40409-017-0138-3

[49] Schliwa M, Fleissner G. The lateral eyes of the scorpion, Androctonus australis. Cell and Tissue Research. 1980;**206**(1):95–114. DOI: 10.1186/s404 09-017-0138-3

[50] Fleissner G. Scorpion lateral eyes: Extremely sensitive receptors of Zeitgeber stimuli. Journal of comparative physiology. 1977;**118** (1):101–108

[51] Hadley NF, Williams SC. Surface activities of some North American scorpions in relation to feeding. Ecology. 1968;**49**:726–734

[52] Prestwich KN. Anaerobic metabolism and maximal running in the scorpion Centruroides hentzi (Banks) (Scorpiones, Buthidae). The Journal of Arachnology. 2006;**34**(2):351–356

[53] Koch LE. A comparative study of the structure, function and adaptation to different habitats of burrows in the scorpion genus Urodacus (Scorpionida, Scorpionidae). Records of the Western Australian Museum. 1978;**6**(2):119–146

[54] Abdel-Nabi IM, McVean A, Abdel-Rahman MA, Omran MA. Intraspecific diversity of morphological characters of the burrowing scorpion Scorpio maurus palmatus (Ehrenberg, 1828) in Egypt (Arachnida: Scorpionida: Scorpionidae). Serket. 2004;**9**(2):41–67

[55] Talal S, Tesler I, Sivan J, Ben-Shlomo R, Tahir HM, Prendini L, Snir S, Gefen E. Scorpion speciation in the Holy Land: multilocus phylogeography corroborates diagnostic differences in morphology and burrowing behavior among Scorpio subspecies and justifies recognition as phylogenandic, ecological and biological species. Molecular Phylogenetics and Evolution. 2015;**91**:226–237. DOI: 10.1016/j.ympev. 2015.04.028

[56] Hadley NF. Environmental physiology. In: Polis GA, editor. The Biology of Scorpions. Stanford, California: Stanford University Press; 1990. p. 321–340

[57] Gibbs AG, Rajpurohit S AG, Rajpurohit S. Cuticular lipids and water balance. In: Blomquist GJ, Bagnères AG, editors. Insect Hydrocarbons: Biology, Biochemistry and Chemical Ecology. New York, NY: Cambridge University Press; 2010. p. 100–120.

[58] Gefen E, Talal S, Brendzel O, Dror A, Fishman A. Variation in quantity and composition of cuticular hydrocarbons in the scorpion Buthus occitanus (Buthidae) in response to acute exposure to desiccation stress. Comparative Biochemistry and Physiology Part A: Molecular & Integrative Physiology. 2015;**182**:58–63

[59] Toolson EC, Hadley NF EC, Hadley NF. Cuticular permeability and epicuticular lipid composition in two Arizona vejovid scorpions. Physiological Zoology. 1977;**50**(4):323–330

[60] Gibbs A, Pomonis JG. Physical properties of insect cuticular hydrocarbons: the effects of chain length, methyl-branching and unsaturation. Comparative Biochemistry and Physiology Part B: Biochemistry and Molecular Biology. 1995;**112**(2):243–249

[61] Dejours P, Ar A. Temperature and starvation affect the hemolymph acid-base balance of the xeric yellow scorpion, Leiurus quinquestriatus. Journal of Comparative Physiology B. 1991;**161**(4):407–412

[62] Gefen E, Ung C, Gibbs AG. C, Gibbs AG. Partitioning of transpiratory water loss of the desert scorpion, Hadrurus arizonensis (Iuridae). Journal of Insect Physiology. 2009;**55**(6):544–548

[63] Gefen E. The relative importance of respiratory water loss in scorpions is correlated with species habitat type and activity pattern. Physiological and Biochemical Zoology. 2011;**84**(1):68–76

[64] Gefen E, Ar AE, Ar A. The effect of desiccation on water management and compartmentalisation in scorpions: the hepatopancreas as a water reservoir. Journal of experimental biology. 2005;**208**(10):1887–1894

[65] Schmidt-Nielsen K. Animal Physiology: Adaptation and Environment. Cambridge: Cambridge University Press; 1997

[66] Warburg MR, Goldenberg S, Ben-Horin A. Thermal effect on evaporative water loss and haemolymph osmolarity in scorpions at low and high humidities. Comparative Biochemistry and Physiology Part A: Physiology. 1980;**67**(1):47–57

[67] Withers PC, Smith GT. Effect of temperature on the metabolic rate and evaporative water loss of the scorpion Urodacus armatus. Journal of Thermal Biology. 1993;**18**(1):13–18

[68] Warburg MR, Ben-Horin A. The response to temperature gradients of scorpions from mesic and xeric habitats. Comparative Biochemistry and Physiology, A. 1981;**68**:277–279

[69] McReynolds CN. Temporal patterns in microhabitat use for the scorpion Centruroides vittatus (Scorpiones: Buthidae). Euscorpius. 2004;**17**:35–45

[70] McReynolds CN. Microhabitat preferences for the errant scorpion, Centruroides vittatus (Scorpiones, Buthidae). Journal of Arachnology. 2008;**36**:557–564

[71] Gefen E, Ar A. Temperature dependence of water loss rates in scorpions and its effect on the distribution of Buthotus judaicus (Buthidae) in Israel. Comparative Biochemistry and Physiology A. 2006;**144**:58–62

[72] Carlson BE, Rowe MP. Temperature and desiccation effects on the antipredator behavior of Centruroides vittatus (Scorpiones: Buthidae). The Journal of Arachnology. 2009;**37**(3):321–330

[73] Inceoglu B, Lango J, Jing J, Chen L, Doymaz F, Pessah IN, Hammock BD. One scorpion, two venoms: prevenom of Parabuthus transvaalicus acts as an alternative type of venom with distinct mechanism of action. Proceedings of the National Academy of Sciences. 2003;**100**(3):922–927

[74] Sridhara S, Chakravarthy AK, Kalarani V, Reddy DC. Diversity and Ecology of Scorpions: Evolutionary Success Through Venom. In: Chakravarthy A, Sridhara S, editors. Arthropod Diversity and Conservation in the Tropics and Sub-tropics. Singapore: Springer; 2016. p. 57–80

[75] Polis GA, McCormick SJ. Scorpions, spiders and solpugids: predation and competition among distantly related taxa. Oecologia. 1986;**71**(1):111–116

[76] Bradley RA. The influence of weather and biotic factors on the behaviour of the scorpion (Paruroctonus utahensis). Journal of Animal Ecology. 1988;**57**:533–551

[77] Yamashita T. Surface activity, biomass, and phenology of the striped scorpion, Centruroides vittatus (Buthidae) in Arkansas, USA. Euscorpius. 2004;**17**:25–33

[78] Skutelsky O. Predation risk and state-dependent foraging in scorpions: effects of moonlight on foraging in the scorpion Buthus occitanus. Animal Behaviour. 1996;**52**:49–57

[79] Polis GA, Myers C, Quinlan M. Burrowing biology and spatial distribution of desert scorpions. Journal of Arid Environments. 1986;**10** (2):137–146

[80] Polis G, Sissom W. Life History. In: Polis GA, editor. The Biology of Scorpions. Stanford: Stanford University Press CA; 1990. p. 161–223

[81] Polis GA. Population and community ecology of desert scorpions. In: Brownell P, Polis GA, editors. Scorpion Biology and Research. Oxford: Oxford University Press; 2001. p. 302–316.

[82] Warburg MR. Intra-and interspecific cohabitation of scorpions in the field and the effect of density,

food, and shelter on their interactions. Journal of Ethology. 2000;**18**(1):59–63

[83] Piñero FS, Tenorio FU, García FJM. Foraging of Buthus occitanus (Scorpiones: Buthidae) on shrub branches in an arid area of southeastern Spain. The Journal of Arachnology. 2013;**41**(1):88–90

[85] Prendini L. Phylogeny and classification of the superfamily Scorpionoidea Latreille 1802 (Chelicerata, Scorpiones): An exemplar approach. Cladistics. 2000;**16**(1):1–78

[85] Lira AF, Foerster SÍ, Badry A. Living in a desert: examining scorpion beta diversity in Egyptian drylands from a macroecological perspective. African Zoology. 2023;**58**(1):18–28. DOI: 10.1080/15627020.2023.2188121

[86] Chown SL, Gaston KJ, Robinson D. Macrophysiology: large-scale patterns in physiological traits and their ecological implications. Functional ecology. 2004;**18**(2):159–167

Section 3

Bats and Owl Companionship

Chapter 4

The Ecology of Urban Plant-Visiting Bats

Theodore H. Fleming

Abstract

Many species of bats, including insectivores and those that visit flowers or eat fruit, live in urban environments. In this chapter, I discuss the ecology of urban plant-visitors (PVs) that belong to two families: the New World Phyllostomidae (American leaf-nosed bats) and the Old World Pteropodidae (flying foxes and their relatives). Topics that I discuss include the extent to which city environments act as filters regarding which PV bat species actually live in them, where they roost, and what they eat. In both families, foliage-roosting and fruit-eating PV bats are most likely to live in urban areas. Their diets often include a mixture of native and exotic plant species. I discuss in detail how four PV bat species live in cities, including the Australian Grey-headed flying fox that now lives in two cities outside the range of its native food plants. Many bats, including some PV species, harbor viruses that can be transmitted to humans. Having these bats living in urban areas, therefore, can pose a health risk to humans, especially in Old World tropical areas where bats are often hunted and eaten.

Keywords: Phyllostomid bats, Pteropodid bats, urban bat diets, urban bat roosts, challenges to urban living for bats, urban bats as disease reservoirs

1. Introduction

Many kinds of bats (Mammalia, Chiroptera) have lived in close association with humans for millennia. Insect-eating species (i.e., members of 18 of 20 recognized bat families) are well-known for roosting and feeding in many urban settings [1–4]. The final two families of bats – American leaf-nosed bats (family Phyllostomidae) and so-called flying foxes of the Old World (family Pteropodidae) – contain many species of plant-visiting bats that pollinate flowers and disperse seeds when they feed on nectar (and pollen) and fruit, respectively. In this chapter, I will discuss how these bats, like their insectivorous relatives, have adapted to urban life. I note that "adaptation" to urban life in these bats probably does not involve evolutionary changes specifically caused by urban features. Instead, it involves a filtering process wherein bats with broad ecological tolerances are attracted to and can survive in cities, whereas species with lower tolerances cannot. Often, the absence of appropriate food resources and/or roost sites serves as a barrier to many of these species in urban areas.

Before looking at urban-dwelling plant-visiting bats in detail, let's consider the two major groups of plant-visiting species and some of their food plants. These

IntechOpen

groups include members of two families – Phyllostomidae and Pteropodidae. The New World family Phyllostomidae (American leaf-nosed bats) contains a total of about 216 species with a diverse array of diets [5]. Its plant-visitors have evolved from insectivorous relatives and belong to four morphologically advanced subfamilies [6]. Nectar feeders occur in two subfamilies (Glossophaginae and Lonchophyllinae) that include about 56 species; frugivores also occur primarily in two subfamilies (Carolliinae and Stenodermatinae) that include about 114 species. Most phyllostomid nectar-feeders visit flowers produced by shrubs and vines, but also some ecologically important trees such as silk cotton (*Ceiba pentandra*) and balsa (*Ochroma lagopus*) in the Malvaceae (subfamily Bombacoideae) and West Indian locust (*Hymenaea courbaril*, Fabaceae). The frugivores include understory feeders (in the Carolliinae) focused on fruits produced by early successional shrubs and trees, especially those produced by *Piper* (Piperaceae) and *Solanum* (Solanaceae), and canopy feeders (in the Stenodermatinae) that consume fruits produced by trees, especially figs (*Ficus* species, Moraceae), introduced mangoes (*Mangifera indica*, Anacardiaceae), and many other native species [7]. Whenever these kinds of flowers and fruits are found in urban areas, they are likely to attract plant-visiting phyllostomids.

In contrast to the Phyllostomidae, members of the Old World family Pteropodidae (flying foxes and their relatives) are strictly plant-visitors [8]. Most of its 201 species feed primarily on fruits that are produced by canopy trees; 14 species that occur in four subfamilies are morphologically specialized to feed at flowers produced by shrubs and trees [9]. Many of the fruit eaters also feed opportunistically on flowers produced by trees. Important fruit sources for these bats include figs, mangoes, parasol leaf trees (*Macaranga grandifolia* and its relatives, Euphorbiaceae), and bananas (*Musa* species, Musaceae). Important flower sources include canopy trees such as *Eucalyptus* and its relatives (Myrtaceae) and *Banksia* shrubs (Proteaceae) in Australia; baobab trees (*Adansonia digitata*, Malvaceae, Bombacoideae), sausage trees (*Kigelia africana*, also Malvaceae), and African tulip trees (*Spathodea campanulata*, Bignoniaceae) in Africa. In Southeast Asia, they visit flowers of the Indian trumpet tree (*Oroxylum indicum*, Bignoniaceae), durian (*Durio zibethinus*, Malvaceae), petai tree (*Parkia speciosa* and its relatives, Fabaceae), and bananas. Only about 15 of its species (8%) are known to live in urban areas [10].

2. The ecology of urban plant-visiting bats

All of these plant-visiting bats live in the tropics and subtropics. Most of them live in tree-dominated habitats (e.g., forests, shrublands, or deserts dominated by large columnar cacti). These are the ancestral homes in which they have evolved over the past 20 million years or so. Urban or suburban environments and fruit plantations are thus relatively new habitats in their evolutionary histories and differ strongly in many features that affect their ecology, behavior, and physiology [4]. These features include ambient temperatures (i.e., the heat island effect), air pollution, light pollution, noise pollution, water quality, predator pressure, and the availability of food resources. Based on these effects, an important question arises: are urban habitats sinks, sources, or neutral for populations of bats compared with population trends in their native habitats? For example, are their mortality rates higher, lower, or the same in urban habitats compared with non-urban habitats?

Another important aspect of urban areas is that they undoubtedly serve as filters that determine which species of plant-visitors are able (or not) to thrive in these new habitats. Important environmental features that act as filters in rural-to-urban gradients include the degree of forest cover, availability of suitable roosting sites, and feeding areas [2, 4]. One obvious expectation is that species with very specific food and shelter needs are less likely to thrive in urban environments than less-specialized (perhaps more 'weedy') species, as is the case in many other groups of animals. Behavioral flexibility is probably an important trait in urban-dwelling bats [1]. So the question becomes: how many phyllostomid and pteropodid species are regular urban-dwellers? How strong is the urban filter effect in these two families?

For phyllostomids, Ramírez-Mejía et al. [11] sampled their species richness and morphological diversity in a transect from forest to urban study sites centered on the city of Villavicencio in lowland Colombia. They reported that species richness declined from a maximum of 18 to 9 as they moved from forest sites into the city. Maximum bat abundance also declined from forests to the city. Missing from the urban samples were foliage-gleaning insectivores, vampires, and very small frugivores. The urban phyllostomid assemblage consisted of large, wide-ranging frugivores (e.g., large species of *Artibeus*), omnivores (e.g., one species of *Carollia*), and one common omnivore-nectar-feeder (a species of *Glossophaga*). Each of these taxa is abundant and widely distributed throughout the lowland American tropics [5]. The results of this study, therefore, support the hypothesis that urban-dwelling phyllostomids are mostly "weedy" species and that the urban filter is strong in these bats.

Occurrence in cities by pteropodid bats has been best-studied in Australia, whose pteropodid fauna contains only eight species: four species of flying foxes (*Pteropus alecto, Pteropus poliocephalus, Pteropus scapulatus,* and *Pteropus conspicillatus*) and single species of *Dobsonia*, *Nyctimene*, *Syconycteris*, and *Macroglossus* [10]. All species are forest-dwellers, but only three species of *Pteropus* also commonly live in cities along Australia's east coast [12]. The other five species are strictly forest dwellers, so the absence of large areas of intact forests in urban areas can be a strong filter for these bats. Compared with the other four species, *Pteropus* bats are generally large (maximum weights of 600 to 1,000 g), usually very abundant (their roosting "camps" can contain 100,000 or more individuals), and they are often seasonally nomadic in the search for flowering trees and fruiting fig trees [13, 14]. In southeastern Queensland and northeastern New South Wales, many camps of three species of *Pteropus* (*P. poliocephalus*, *P. alecto*, and *P. scapulatus*) occur in both urban and non-urban areas [15]. *P. conspicillatus* occurs only in northeastern Queensland, where the density of urban areas is much lower. Thus, like certain phyllostomids, only a subset of Australia's plant-visiting bats routinely live in urban areas.

Pteropodids also occur in Africa and Asia, where they sometimes live in or near urban areas. In Africa, the strong-flying and wide-ranging straw-colored fruit bat *Eidolon helvum,* weighing 250 to 311g is the most common urban bat throughout the continent [8, 10]. It roosts throughout this region in tall trees, often in residential areas, and its colonies sometimes contain hundreds of thousands of bats. Its diet consists mainly of fruit (e.g., 31 species in 16 families, including many figs, in Kenya), and it also visits flowers of the silk-cotton tree (*Ceiba pentandra*) [10, 16, 17]. It flies distances of 30 km or more to feeding areas each night and seasonally migrates hundreds of kilometers to food-rich habitats north and south of the forests of central Africa [16, 18–20].

In Southeast Asia, two foraging guilds of pteropodids occur in southern Thailand and elsewhere. These guilds include the nectarivores *Eonycteris spelaea, Macroglossus sobrinus*, and *M. minimus,* and frugivores, including three species of *Cynopterus* (*C. brachyotis*, *C. sphinx*, and *C. horsfieldii*) and *Rousettus leschenaultii.* Known urban dwellers among these species include *C. brachyotis* and *C. sphinx,* whereas the other species live in forests and plantations [21]. An analysis of habitat use by these species in southern Thailand indicated that the abundance of frugivores increased near urban areas, whereas the abundance of nectarivores decreased [21]. These results again indicate that urban areas are acting as strong filters on habitat use by pteropodid bats and that frugivores are more tolerant of human disturbance than nectarivores. A detailed analysis of the diet of the common *C. brachyotis* in Singapore supports this conclusion [22]. Their results indicated that it fed on the fruits of at least 10 native species and 17 non-native species. Unlike other nectar-feeding pteropodids, the cave-dwelling nectarivore *E. spelaea* lives in an urban cave near Kuala Lumpur in Peninsular Malaysia and visits the flowers of at least 24 and 16 species of native and exotic plants, respectively. These results provide an important insight into why this species and other fruit-eating pteropodids can thrive in urban settings.

Crucial requirements for all bats include sources of food and shelter. For the plant-visitors just discussed, the saying "If it's planted, they will come" is likely to be true. Since many of the above-mentioned plants are commonly grown in urban settings, these food sources and many other native and exotic species will often attract plant-visiting bats. In addition, predator-free and physiologically suitable day roosts are also critical for them. In terms of roost use, members of the two plant-visiting families are quite different. For example, because they can echolocate, phyllostomid fruit-eaters and nectar-feeders often roost in dark caves, hollow trees, and mines [23]. They also roost in culverts and abandoned buildings. In addition to caves, members of the fig-eating Stenodermatinae often roost in foliage, either in forest understories or tree canopies. In contrast, pteropodid bats cannot echolocate and hence are not cave-roosters (with a few exceptions, e.g., *E. spelea*). Most of them are therefore foliage roosters in forest understories and tree canopies [8].

Differences in the roosting behavior of these two families can lead to significant differences in the way they interact with urban environments. For example, the day roosts of foliage-roosting bats are often located within urban areas, especially in parks, whereas the day roosts of bats living in caves or mines are not. As a result, the foraging behavior can differ substantially between these two groups of bats. Cave roosters often have to travel substantial distances to feed in urban areas, whereas foliage-roosters theoretically do not. I use the word "theoretically" because some pteropodid bats roost in urban areas but fly substantial distances to feed away from them [13, 14]. They apparently view the benefits of roosting in cities as worth the costs of long-distance commutes.

3. Examples of urban-dwelling, plant-visiting bats

To illustrate how some well-studied plant-visiting bats thrive in urban areas, I will discuss the following species in detail: the phyllostomids *Artibeus lituratus* (a frugivore) and *Leptonycteris yerbabueanae* (a nectarivore), as well as the pteropodids *C. brachyotis* and *P. poliocephalus* (both primarily frugivores that also visit flowers opportunistically; see **Figure 1**).

Figure 1.
Four plant-visiting bats that live in urban areas. Two phyllostomid bats include Artibeus lituratus *(a) and* Leptonycteris yerbabuenae *(b). Two pteropodid bats include* Cynopterus brachyotis *(c) and* Pteropus poliocephalus *(d). Source of images: Wikimedia Images.*

The so-called Great fruit-eating bat *Artibeus lituratus* (see **Figure 1a**) is widely distributed throughout the lowland Neotropics, from northern Mexico to southern Brazil. Weighing 53–73 g, it is the largest fruit-eating phyllostomid and is a member of the subfamily of fig-eating specialists, Stenodermatinae [8]. Its use of habitats is broad but is concentrated in wet to moist lowland tropical forests. In Venezuela, for example, Handley [24] reported capturing few individuals in dry habitats. While principally a fig-eater (*Ficus*), its diet is broad and contains at least 146 species in 35 families throughout its geographic range [7]. In particular habitats, however, its diet is known to include only about 10–13 fruit species. Most of these are produced by canopy trees rather than understory shrubs. It also occasionally eats the leaves of these trees [25]. Like many other frugivorous phyllostomids, *A. lituratus* has a bimodal, polyestrous breeding system in which adult females produce a single pup twice a year [26].

Its roosts include the foliage of many kinds of canopy trees, the thatch of palm trees, and, infrequently, porches or wooden canopies of urban buildings [27–29]. Roosts are occupied by small groups of bats, usually including a single adult male, several adult females, and their offspring. As a result, its social system is based on harem-polygyny [30].

Given its predilection for eating figs, whose trees are commonly planted in urban and suburban areas, and its preference for roosting in foliage rather than in caves, this common bat is clearly pre-adapted for living in human-dominated systems. Both its food and potential roost sites are readily available in these systems.

Disturbances or threats in such areas include ambient temperatures, light pollution, and predators. For example, careful examination of the behavior of a group of *A. lituratus* roosting in an urban palm tree revealed that their general activity, including self-grooming, increased as ambient temperatures increased, and their rest times decreased [31]. The long-term demographic consequences of rising or high urban temperatures for this foliage-roosting bat, however, are currently unknown.

Compared with other common frugivorous phyllostomids (e.g., *Carollia perspicillata* and *Sturnira lilium*), *A. lituratus*'s foraging activity is not particularly sensitive to moonlight levels [32], but observations in Mérida, Venezuela, indicate that it is sensitive to light pollution. It avoids areas of strong lighting when foraging [33]. Regarding predation, it is known to be eaten by the bat falcon (*Falco rufigularis*) in urban areas [34], and although it has not yet been documented [35], arboreal boiid snakes are likely to consume canopy-roosting bats. Overall, however, the benefits of living in urban areas likely far outweigh the costs for this species, and as a result, it is a common bat in human habitats.

The second phyllostomid that I will discuss is the Lesser long-nosed bat, *Leptonycteris yerbabuenae* (see **Figure 1b**), which is primarily a nectar-feeding species. Weighing 23–30g, it is the second-largest member of its subfamily (Glossophaginae). Occurring in arid habitats from southern Arizona to northern Central America, it is one of the northernmost members of its entire family [36]. It is strongly migratory, with both sexes spending the winter in tropical dry forests in southern Mexico and northern Central America, where they feed at the flowers of canopy trees [37]. In spring, many females, now pregnant with a single pup, migrate up to 1,200 km north to give birth in maternity roosts located in northwestern Mexico and southwestern Arizona, mainly in the Sonoran Desert. During much of this period, they feed heavily on the flowers and fruits of tall columnar cacti such as cardón (*Pachycereus pringlei*) and saguaro (*Carnegiea gigantea*) and the flowers of century plants (*Agave* species) [38].

Unlike most phyllostomids, *L. yerbabuenae* is a highly gregarious bat. Its main roosts are warm caves and mines that can contain many thousands (up to 100,000 or more) of bats. Because they are built for efficient, long-distance flight [39], many individuals feed tens of kilometers (up to almost 100 km) away from their day roosts each night [40, 41]. This foraging method allows them to roost in large caves located some distance from cities and still forage in rural and urban areas each night whenever appropriate food plants (mostly columnar cacti and agaves) are available [42].

In southern Arizona, *L. yerbabuenae* roosts during the day in mountain caves and mines and feeds at hummingbird feeders in rural and urban areas. They often fly tens of kilometers one-way to do this [43, 44]. Their feeding behavior in the Tucson area has been monitored since 2007 by participants in a citizen-science project [45]. Results of this project indicate that bats visit these feeders for about two months each year (mid-August to late-October) before migrating back into Mexico; that the number of bats visiting feeders in Tucson has increased since 2007 and now involves thousands of bats; and that most of the bats visiting urban feeders are either juveniles or yearlings and not adults, which feed mostly at *Agave* flowers in far southeastern Arizona. Recent work with radio-tagged bats in Tucson indicates that many of them are highly consistent in visiting particular feeders each night [46].

Leptonycteris yerbabuenae is an example of a bat that is not specifically adapted to living full-time in urban settings but whose foraging behavior is flexible enough to take advantage of an "unnatural" food source (hummingbird feeders) when it is available. In this sense, urban areas throughout its range can potentially (and actually, in southern Arizona) increase the carrying capacity for this and other species of nectar-feeding bats. As long as safe and physiologically appropriate (i.e., warm rather than cool) roosts are available within the foraging range of these bats, urban areas can thus have a positive effect on their populations.

Let's now shift our attention to Old World plant-visiting bats that sometimes live and feed in urban areas. This includes bats of the genus *Cynopterus* (Pteropodidae; see **Figure 1c**). Two species, the Lesser short-nosed fruit bat (*C. brachyotis*) and the Greater short-nosed fruit bat (*C. sphinx*), are common forest dwellers throughout South India and Southeast Asia [10]. They also often reside in urban and other human-created areas such as fruit farms, tea plantations, and oil palm plantations in Asia [21, 47]. As their name implies, these two species differ somewhat in size; *C. brachyotis* weighs 21–32 g, and *C. sphinx* weighs 53–75 g. In many respects, these bats are the Old World ecological analogues of New World large fig-eating bats (e.g., *Artibeus lituratus* and *A. jamaicensis*).

Diets of these bats include fruits, leaves, and flowers. The diet of *C. brachyotis* in Malaysia includes at least 54 species of fruit from 14 families, 14 species of leaves (mostly from the family Fabaceae), and 4 species of flowers, also from Fabaceae (formerly Leguminosae) [22, 48]. Because they are common, available all year, and a good energy source, figs (*Ficus*) are a main component of its diet. Legume leaves are an important source of protein for these bats. The diet of city-dwelling *C. brachyotis* in Singapore includes 33 species of fruit in 21 families [22]. These species include 10 native plants and 17 exotic plants, which supports the idea that urban-dwelling probably increases rather than decreases its diet breadth. The three most common fruit species in the samples were native plants – *Fagraea fragrans* (Gentianaceae), a species of *Ficus*, and *Terminalia catappa* (Combretaceae). A study of the diet and foraging behavior of *C. brachyotis* in disturbed rainforests in Malaysia indicated that bats traveled about 295 m from their day roosts to feed, mostly on *Ficus* fruits, in small home ranges that averaged about 3.2 ha in size [22]. Thus, in native forests, this species appears to be a short-distance forager, primarily because its foliage roosts are often located near high densities of fruiting *Ficus* trees. Its foraging behavior in urban areas is likely to be similar to this.

Like their New World counterparts, these bats are foliage roosters. In Malaysia, roost use in *C. brachyotis* differs from that of *C. sphinx* in that it appears to use the leafy foliage of non-palm trees. In contrast, *C. sphinx* is apparently unique in Asia for creating its own roosts by clipping large, often palm, leaves with their teeth to make "tents" [21, 49]. In *C. sphinx* in western India, adult males are the tent-makers, and tents are often seasonally occupied. Their inhabitants include one adult male and 1–10 adult females, and their mating system, like that of *A. lituratus*, involves harem-polygyny [50]. In addition, like that of *A. lituratus*, its reproductive cycle involves bimodal polyestry, with females giving birth to a single pup twice a year.

Cynopterus sphinx has been studied in several urban areas. In Hong Kong, it commonly roosts in urban parks and constructs tents mainly in tall palm trees. Its diet there includes at least eight fruit species, including five species of *Ficus*, and the number of fruit trees in a park, rather than its area, predicts the number of

C. sphinx roosts in a park [51, 52]. Many roosts are located near fruit resources, so it appears that this bat, like *C. brachyotis* in Malaysia, is a short-distance forager. In a review of the effect of urbanization on *C. sphinx* in India, Garg et al. [53] noted that harem size and reproductive output are lower in a city compared with a nearby rural area, suggesting that its demography is sensitive to urban conditions.

Microclimate temperature and artificial lighting are also known to affect the roosting and foraging behavior of urban populations of *C. sphinx*. For example, in the Lucknow region of northeastern India, Kumar et al. [54] monitored the temperatures occurring in roosts in palm trees and an abandoned house where these bats were roosting. They found that as daytime temperatures rose to over 45°C, bats moved from foliage tents to cooler microclimates in abandoned buildings, an infrequently used roost type by this species. These results suggest that urban-dwelling, foliage-roosting bats (probably in both the Old and New Worlds) will suffer from rising temperatures associated with urban heat islands and climate change. Light pollution also affects the foraging behavior of *C. sphinx* in the Lucknow region [54]. These researchers reported that feeding visits to brightly illuminated fruit trees were much lower than visits to dimly lit or dark fruit trees.

In summary, *Cynopterus* fruit bats roost in a variety of habitats throughout Asia, including urban areas. They are important dispersers of seeds produced by a variety of native and exotic species. In urban areas in northeastern India, at least, their roosting and foraging behavior appears to be sensitive to daytime roost temperatures and nocturnal light pollution. While urban areas provide many roosting and feeding opportunities, their microclimates and lighting conditions are likely to be detrimental to these species. Therefore, the continuing growth of urban areas, combined with climate change, will probably have negative effects on these and other foliage-roosting bats in the future.

The final species whose roosting and foraging behaviors I will discuss is the Grey-headed flying fox (*Pteropus poliocephalus*; see **Figure 1d**) which lives along the east coast of Australia. Weighing 600–1,000g and with a wingspan of about 1.2 m, this is the largest bat in Australia [12]. It is highly gregarious and often lives in large colonies (camps) in the foliage of trees, where they are exposed to high daily temperatures and intense sunlight. Its camps can contain tens of thousands of individuals. Historically, these camps have been divided into "summer" (September to April) and "winter" (April to September) roosts [14]. Most of its reproductive activity occurs in summer camps, with adult males forming territories or harems containing a few females. Mating occurs mostly in April, and the birth of a single pup per female takes place in October. Occupancy of many of these camps changed seasonally in response to changes in the availability of flowers and fruit [13, 55, 56]. In recent times, however, occupancy of many camps has not changed seasonally [15].

The diet of *P. poliocephalus* includes a sequence of flowers followed by fruit as the reproductive phenology of different species changes seasonally [57]. Its flower diet comes from at least six native families, including those produced by eucalypts and their relatives (Myrtaceae). Its fruit diet includes at least nine native families and is often concentrated on figs (*Ficus*, Moraceae) [10]. When foraging, these bats often fly 20–50 km from their day roosts, so, unlike many smaller pteropodids, they are relatively long-distance foragers that can provide substantial movement for pollen and seeds [58]. As a result, this species is an important "mobile link" between many plant populations, both within urban and non-urban areas.

Camps of *P. poliocephalus* are now often located in or near major urban areas, including Adelaide, Melbourne, Sydney, and Brisbane. Roost surveys conducted over 15 years in North Queensland, for example, indicate that these bats are becoming more urbanized there [59, 60]. Results in Queensland indicate that this trend results from roost choices made by these bats rather than from the loss of native habitat. In general, urban roosts often differ from non-urban ones by being smaller; their bats are less nomadic because of the year-round availability of food resources, including many non-native species, in cities [15].

In the last few decades, *P. poliocephalus* has moved into more southern urban areas. In 2011, this bat established a camp in Adelaide in its Botanic Park. The initial size of the colony was estimated to be about 300 bats, but it had increased to about 20,000 bats by 2020 [59]. These bats fed heavily on several species of introduced plants, including a close relative of *Eucalyptus* and a fig. Owing to the high density of non-native food sources in this city and its suburbs, five satellite-tagged bats traveled relatively short distances that averaged about 4 km from the roost to feed in the spring. In 1986, *P. poliocephalus* formed a camp in the Royal Botanic Garden in Melbourne. Since then, the size of this camp has been increasing. Its diet there is concentrated on two kinds of resources that are not native to Melbourne: flowers of 22 species of *Eucalyptus* and fruits of eight species of *Ficus* [61]. Both of these examples provide dramatic support for the idea that "If you plant it, they will come."

4. Urban plant-visiting bats as disease reservoirs

Each of these four and other tropical plant-visiting bats that have become quite urbanized must contend with a variety of physical challenges, including heat island effects, light and other kinds of pollution, and new predators in cities. In addition, becoming urbanized sometimes places these animals in direct conflict with humans. Large roosts of pteropodids, for example, are noisy and smelly and, hence, are often not appreciated or tolerated by humans. Another important source of conflict results from the fact that many species of bats, including those that live in cities, harbor many kinds of viruses and bacteria. Infamous examples of these viruses include rabies, COVID-19, and SARS-CoV-2, and both bats and birds are known to harbor many other kinds of pathogens [62–64]. Their ability to serve as reservoirs for many pathogens without suffering ill-effects has recently resulted in much scientific attention [65].

When bats live in urban environments, they can transmit viruses to humans by at least two different routes. The first route is direct transmission whenever bats are killed and eaten. Pteropodid and other bats throughout Asia and the South Pacific often occur in traditional diets, and hence the possibility of direct disease transmission can be high [66, 67]. The second route is when bats transmit viruses to other animals (domestic or otherwise) that come in contact with humans, again often in their diets. Here, I will discuss briefly three examples of virus spillover to humans that occur in the Old World tropics and subtropics of Africa, Asia, and Australia. In these regions, bats are known to be infected with a number of viruses that can cause deaths in humans. These include Ebola, SARS-Cov1 and 2, Nipah, and Hendra.

4.1 The Ebola virus in Africa

This virus is an important source of human deaths in Africa. Through 2016, over 11,000 deaths from Ebola infection have been recorded [68]. At least five species of pteropodid bats – *Eidolon helvum*, *Hypsignathus monstrosus*, *Epomops franqueti*, *Myonycteris torquata*, and *Rousettus aegyptiacus* – are suspected to be reservoirs of Ebola in Central and West Africa [69, 70]. Various African primates, including gorillas and chimpanzees, are also known to harbor Ebola [71, 72]. Transmission of this virus to humans usually occurs in rural areas as a result of contact with live bats and primates and their fluids, as well as consumption of dead animals. Increased contact with infected wildlife likely happens as a result of the usual causes: habitat destruction and human population growth. Human-to-human transmission in both rural and urban areas can also occur when infected individuals and their fluids come into close contact with other individuals.

4.2 SARS in Asia

Asian bats known to harbor SARSr-CoV include members of the insectivorous families Hipposideridae, Rhinolophidae, Molossidae, Vespertilionidae, and the plant-visiting Pteropodidae [73, 74]. In Asia, these bats inhabit a large area encompassing nearly 500 million people. Of these people, about 39,000 are infected with SARS annually as a result of direct human-bat or other mammal contact. Infection rates are likely similar in rural and urban settings due to the transport of dead bats and other mammals from rural areas to urban areas for consumption. Other mammals harvested for food and infected with SARS include civet cats, raccoon dogs, and livestock such as cows, pigs, and rabbits, so bats are not the only source of SARS infection in Asian human populations.

4.3 Hendra virus in Australia

In Australia, unlike Asia, spillover of corona viruses from bats to humans has apparently not yet occurred [75, 76]. In contrast, Eby et al. [77] reviewed the spillover of Hendra virus to humans there, either directly or through an intermediate host such as horses, from pteropodid bats over a 25-year period. Two major factors – recent land use changes (i.e., conversion of natural habitats to human-dominated ones) and natural food shortages that cause pteropodids to search for food in agricultural and urban areas – lie behind the spillover of this virus to humans. Winter food levels undergo occasional decreases, forcing bats to rely more heavily on non-native food species, and this is when the occurrence of shedding of Hendra viruses and infection of humans is greatest.

These and many other studies clearly indicate that as contact with bats has increased, the transmission of various viral and other diseases to humans has also increased. This increase has occurred for several reasons, including an increase in human-dominated systems (e.g., urban and agricultural areas), a decrease in intact natural areas, and the planting of exotic plant species that can serve as food for plant-visiting bats in human-dominated areas, as well as human population growth. Whenever food and safe roost sites occur in human-dominated sites, some bats are likely to take advantage of them and can pose a health risk to humans, especially when bats are killed for food. In the New World tropics, the occurrence of rabies, particularly in vampire bats (usually the phyllostomid Common Vampire (*Desmodus rotundus*)), has increased as a result of the occurrence of large populations of domestic animals that are often their major food source [77–79]. Whenever acceptable fruit and flower resources are planted and livestock is raised in human-dominated areas, the carrying capacity for certain species of bats has increased. If these species are reservoirs for disease organisms, then their transmission rates to humans will likely also increase.

5. Conclusions

Like many insect-eating bats, a number of plant-visiting bats in the families Phyllostomidae and Pteropodidae live in urban and other human-dominated habitats. Studies have shown that only certain plant-visiting species have adopted an urban lifestyle. These bats are usually widely distributed, abundant, and flexible regarding their food and roost requirements. They also usually live in foliage roosts rather than in caves and are primarily fruit-eaters. The planting of native and exotic fruit and bat-pollinated trees often provides these bats with a year-round source of food and thus makes urban areas attractive places to live. In these habitats, bats must deal with increased daytime roost temperatures and light pollution, but nonetheless, they often thrive in urban areas. An important reason for this is that they don't have to forage as far from their day roosts compared with bats living in native habitats. In urban-dwelling Australian pteropodids, roost sizes tend to be smaller than roosts in native habitats owing to the higher food densities in cities. The demographic consequences of urban life for plant-visiting bats compared with bats living in native habitats are largely unknown but are likely to be different. For example, as in other vertebrates, the timing of mating and births is likely to be different. There is also evidence that the reproductive output of urban bats may be lower in cities than away from them, at least in Australian pteropodids [53]. Finally, because many bat populations are reservoirs for viruses that can cause human fatalities, urban bats can sometimes pose a health threat to humans. Nonetheless, because they are important dispersers of pollen and seeds of many native and exotic plants, urban plant-visiting bats also provide important positive ecosystem services wherever they live, including urban areas.

Acknowledgments

Many colleagues, students, field assistants, and volunteers have helped me during my decades-long studies of bat ecology in Latin America, Australia, and the United States. Support for these studies has been provided by the Universities of Missouri-St. Louis and Miami, as well as various funding agencies, including the U.S. National Science Foundation, the National Geographic Society, and the Fulbright Foundation. I especially thank Drs. Danilo Russo and Romeo Saldaña-Vásquez for providing me with copies of their important review papers for this chapter. Mil gracias a todos!

Author details

Theodore H. Fleming
University of Miami, Coral Gables, USA

*Address all correspondence to: flemingt383@gmail.com

IntechOpen

© 2025 The Author(s). Licensee IntechOpen. This chapter is distributed under the terms of the Creative Commons Attribution License (http://creativecommons.org/licenses/by/4.0/), which permits unrestricted use, distribution, and reproduction in any medium, provided the original work is properly cited.

References

[1] Jung K, Threlfall CG. Urbanisation and its effects on bats – A global meta-analysis. In: Voigt CC, Kingston T, editors. Bats in the Anthropocene: Conservation of Bats in a Changing World. Switzerland: Springer; 2016. p. 13–33

[2] Russo D, Coleman JL, Ancilloto L, Corine C, et al. Ecosystems services by bats in urban areas. In: Moretto L, editor. Urban Bats. Switzerland: Springer Nature; 2023. p. 167–180. DOI: 10.1007/978-3-031-13173-8_12

[3] Saldaña-Vázquez RA, MacSwiney MC, Bolivar-Cime B, Avila-Flores R, et al. Mexican bats: Threats in the Anthropocene. In: Jones RW, editor. Mexican Fauna in the Anthropocene. Switzerland: Springer Nature; 2023. p. 237–265

[4] Saldaña-Vázquez RA, MacSwiney MC, Bolivar-Cime B, Avila-Flores R, et al. Mexican bats: Threats in the Anthropocene. In: Jones RW, editor. Mexican Fauna in the Anthropocene. Switzerland: Springer Nature; 2023. p. 237–265

[5] Fleming TH, Davalos LM, Mello MAR, editors. Phyllostomid Bats, a Unique Mammalian Radiation. Chicago: University of Chicago Press; 2020

[6] Cirranello AL, Simmons NB. Diversity and discovery: A golden age. In: Fleming TH, Davalos LM, Mello MAR, editors. Phyllostomid Bats, a Unique Mammalian Radiation. Chicago: University of Chicago Press; 2020. p. 43–62

[7] Lobova TA, Geiselman CK, Mori S. Seed Dispersal by Bats in the Neotropics. Bronx: The New York Botanical Garden; 2009

[8] Nowak RM. Walker's Mammals of the World. 5th ed. Baltimore: The Johns Hopkins University Press; 1991

[9] Aziz SA, McConkey KR, Tanalgo K, Sritongchuay T, et al The critical importance of Old World fruit bats for healthy ecosystems and economies. Frontiers in Ecology and Evolution. 2021;**9**:1–29

[10] Mickleburgh SP, Hutson AM, Racey PA, editors. Old World Fruit Bats, an Action Plan for Their Conservation. Gland, Switzerland: International Union for the Conservation of Nature and Natural Resources; 1992

[11] Ramírez-Mejía AF, Urbina-Cardona JN, Sánchez F. Functional diversity of phyllostomid bats in an urban–rural landscape: A scale-dependent analysis. Biotropica. 2020;**2020**:1–15. DOI: 10.1111/btp.12816

[12] Strahan R, Editor. Complete Book of Australian Mammals. Sydney, Angus & Robertson: 1983

[13] Eby P. Seasonal movements of grey-headed flying foxes, Pteropus poliocephalus, from two maternity camps in northern New South Wales. Wildlife Research. 1991;**18**:547–559

[14] Tidemann CR. The biology and management of the grey-headed flying-fox, Pteropus poliocephalus. Acta Chiropterologica. 1998;**1**:151–164

[15] Lunn TJ, Eby P, Brooks R, McCallum H, et al Conventional wisdom on roosting behavior of

Australian flying-foxes— A critical review, and evaluation using new data. Ecology and Evolution. 2021;**2021**:13532–13558

[16] Hurme E, Jakob Fahr J, Network EM, Eric-Moise BF, et al Fruit bat migration matches green wave in seasonal landscapes. Functional Ecology. 2022;**2022**:2043–2055

[17] Richter HV, Cumming GS. Food availability and annual migration of the straw-colored fruit bat (Eidolon helvum). Journal of Zoology. 2006;**268**:35–44. DOI: 10.1111/j.1469-7998.2005.00020.x

[18] Richter HV, Cumming GS. First application of satellite telemetry to track African straw-coloured fruit bat migration. Journal of Zoology. 2008;**275**:172–176. DOI: 10.1111/j. 1469-7998.2008.00425.x

[19] Thomas DW. The annual migration of three species of West African fruit bats (Chiroptera: Pteropodidae). Canadian Journal of Zoology. 1983;**61**:2266–2272

[20] Campbell P, Reid NM, Zubaid A, Adnan AM, Kunz TH. Comparative roosting ecology of Cynopterus (Chiroptera: Pteropodidae) fruit bats in Peninsular Malaysia. Biotropica. 2006;**38**:725–734

[21] Sritongchuay T, Hughes AC, Bumrungsri S. The role of bats in pollination networks is influenced by landscape structure. Global Ecology and Conservation. 2019;**20**:1–13. DOI: 10.1016/j.gecco.2019.e00702

[22] Chan AAQ, Aziz SA, Clare EL, Coleman JL. Diet, ecological role and potential ecosystem services of the fruit bat, Cynopterus brachyotis, in a tropical city. Urban Ecosystems. 2021;**24**: 251–263

[23] Rodriguez-Duran A. Roosting ecology: The importance of detailed decriiption. In: Fleming TH, Dávalos LM, Mello MAR, editors. Phyllostomid Bats. Chicago: University of Chicago Press; 2020. p. 311–324

[24] Handley Jr CO. Mammals of the Smithsonian Venezuelan Project. Brigham Young University Science Bulletin Biological Series. 1976;**20**:1–91

[25] Zortéa M, Mendes SL. Folivory in the Big fruiteating bat, Artibeus lituratus (Chiroptera: Phyllostomidae) in eastern Brazil. Journal of Tropical Ecology. 1993;**9**:117–120. DOI: 10.1017/S0266467400007057

[26] Barclay RMR, Fleming TH. Reproduction and life histories. In: Fleming TH, Davalos LM, Mello MAR, editors. Phyllostomid Bats, a Unique Mammalian Radiation. Chicago: University of Chicago Press; 2020. p. 205–222

[27] Batista CB, Pereira AD, Figueiredo GDT, de Lima IP. See you later: Roost fidelity of the "Boomerang Bats", Artibeus lituratus (Olfers, 1818) (Mammalia, Chiroptera). Ciências Biológicas E da Saúde. 2024;**45**:27–34. DOI: 10.5433/1679-0367.2024v45n1p27

[28] Langlois GD, Stevens RD. Comparison of roosting behavior between two disparate landscapes by a Neotropical bat (Artibeus lituratus) in the Atlantic Forest of Paraguay. Behavioral Ecology and Sociobiology. 2024;**78**:106. DOI: 10.1007/s00265-024-03516-w

[29] Muñoz-Romo M, Herrera EA, Kunz TH. Roosting behavior and group

stability of the big fruit-eating bat Artibeus lituratus (Chiroptera: Phyllostomidae). Mammalian Biology. 2008;**73**:214–221. DOI: 10.1016/j.mambio.2007.05.013

[30] McCracken GF, Wilkinson GS. Bat mating systems. In: Crichton EG, Krutzsch PH, editors. Reproductive Biology of Bats. San Diego: Academic Press; 2000. p. 321–362

[31] Generoso S, Schaucoski VQ, Miranda JMD, Colares R, Carvalho F. Temperature and pups influence daytime roosting behavior of the great fruit-eating bat. Artibeus lituratus, in an Urban Southern Brazilian Habitat. Mammalia 2025;aop. 2025. DOI: 10.1515/mammalia-2024-0100

[32] Mello MAR, Kalko EKV, Silva WR. Effects of moonlight on the capturability of frugivorous phyllostomid bats (Chiroptera: Phyllostomidae) at different time scales. Zoologia. 2013;**30**:397–402. DOI: 10.1590/S1984-46702013000400005

[33] Hoyos-Díaz JM, Villalba-Alemán E, Ramoni-Perazzi P, Muñoz-Romo M. Impact of artificial lighting on capture success in two species of frugivorous bats (Chiroptera, Phyllostomidae) in an urban locality from the Venezuelan Andes. Mastozoologia Neotropical. 2018;**25**:473–478. DOI: 10.31687/saremMN.18.25.2.0.22

[34] Seijas AE, Quintero AA, Salazar G, Arangueran DP. Aves de la Cuidad de Guanare, Portuguesa,Venezuela. Boletin Del Centro de Investigaciones Biologicas. 1996;**45**:55–76

[35] Esperad CEL, Vrcibradic D. Snakes preying on bats: New records from Brazil and a review of recorded cases in the Neotropical region. Revista Brasileira De Zoologia. 2007;**24**:848–853

[36] Cole RF, Wilson DE. Leptonycteris yerbabuenae. Mammalian Species. 2006;**797**:1–7

[37] Arizmendi MD, Valiente-Banuet A, Rojás-Martínez A, Davila-Aranda P. Columnar cacti and the diets of nectar-feeding bats. In: Fleming TH, V-B, editors. Columnar Cacti and Their Mutualists. Tucson, University of Arizona Press; 2002. p. 264–282

[38] Fleming TH, Nassar JM. The population biology of the lesser long-nosed bat Leptonycteris curasoae in Mexico and Northern South America. In: Fleming TH, Valiente-Banuet A, editors. Columnar Cacti and Their Mutualists: Evolution, Ecology, and Conservation. Tucson, University of Arizona Press; 2002. p. 283–305

[39] Sahley CT, Horner MA, Fleming TH. Flight speeds and mechanical power outputs of the nectar-feeding bat, Leptonycteris curasoae (Phyllostomidae: Glossophaginae). Journal of Mammalogy. 1993;**74**:594–600

[40] Goldshtein A, Handel M, Eitan O, Bonstein A, Shaler T, et al Reinforcement learning enables resource partitioning in foraging bats. Current Biology. 2020;**30**:1–7

[41] Horner MA, Fleming TH, Sahley CT. Foraging behaviour and energetics of a nectar-feeding bat, Leptonycteris curasoae (Chiroptera: Phyllostomidae). Journal of Zoology. 1998;**244**:575–586

[42] Brioñes-Salas M, Medina-Cruz GE, Martín-Regalado CN. Taxonomic, functional, and phylogenetic diversity

of bats in urban and suburban environments in southern México. Diversity. 2024;**16**:1–16. DOI: 10.3390/d16090527

[43] Buecher DC, Sidener R 2013. Long distance commutes by lesser long-nosed bats (Leptonycteris yerbabuenae) to visit residential hummingbird feeders. In: Proceedings of the Madrean Conference 2012.

[44] Lowery SF, Blackman ST, Abbate D. Urban movement patterns of lesser long-nosed bats (Leptonycteris curasoae): Management implications for the Habitat Conservation Plan within the city of Tucson and town of Marana. Phoenix, Research Branch, Arizona Game and Fish Department. 2009;**25** 1–25

[45] Fleming TH, Richardson S, Scobie EM. "Free" food: Nectar bats at hummingbird feeders in southern Arizona. Journal of Mammalogy. 2021;**102**:1128–1137. DOI: 10.1093/mammal/gyab054

[46] Krivoruchko K, Namaani G, Yovel Y, Fleming TH. Consistency of use of hummingbird feeders by the nectar-feeding bat Leptonycteris yerbabuenae (Chiroptera, Phyllostomidae) in the Tucson Basin, Arizona. Acta Chiropterologica. 2024;**26**:227–235. DOI: 10.3161/15081109ACC2024.26.2.007

[47] Funakoshi K, Akbar Z. Behavoural and reproductive ecology of the dog-faced fruit bats. Cynopterus Brachyotis and C. Horsefieldi, in a Malaysian Rainforest. Mammal Study. 1997;**22**:95–108

[48] Chan AAQ, Aziz SA, Clare EL, Coleman JL. Diet, ecological role and potential ecosystem services of the fruit bat, Cynopterus brachyotis, in a tropical city. Urban Ecosystems. 2021;**24**: 251–263

[49] Tan KH, Zubaid A, Kunz TH. Food habits of Cynopterus brachyotis (Muller) (Chiroptera: Pteropodidae) in Peninsular Malaysia. Journal of Tropical Ecology. 1998;**14**:299–307

[50] Storz JF, Bhat HR, Kunz TH. Social structure of a polygynous tent-making bat. Cynopterus Sphinx (Megachiroptera). Journal of Zoology. 2000a;**251**:151–165. DOI: 10.1111/j.1469-7998.2000.tb00600.x

[51] Storz JF, Balasingh J, Thiruchenthil NP, Emmanuel K, Kunz TH. Dispersion and site fidelity in a tent-roosting population of the short-nosed fruit bat (Cynopterus Sphinx) in southern India. Journal of Tropical Ecology. 2000b;**16**:117–131

[52] Seeburran G 2019. Habitat Suitability of Hong Kong's Urban Parks for Short-nosed Fruit Bats Cynopterus Sphinx. Hong Kong, University of Hong Kong. M.S. thesis.

[53] Garg KM, Chattopadhyay B, Doss PS, Kumar V, Kandula S, et al. Bat societies across habitat types: Insights from a commonly occurring fruit bat Cynopterus Sphinx. In: Moretto L, Coleman JL, Davy CM, Fenton MB, editors. Urban Bats. Cham: Springer; 2022. p. 61–71

[54] Kumar V, Shankar S, Acharya S, Gautam D, Elangovan V. Seasonal influence on roosting ecology of the Short-nosed fruit bat, Cynopterus Sphinx. Indian Journal of Ecology. 2025;**52**:811–820. DOI: 10.55362/IJE/2025/4579

[55] Kumar V, Elangovan V, Saroj S, Anjum KN. Observations on the effect

of Artificial Lights on foraging behaivour of the Greater short-nosed bat, Cynopterus Sphinx (Vahl, 1797) in Babasaheb Bhimrao Ambedkar University campus, Lucknow. India Records of the Zoological Survey of India. 2024;**124**:547–564. DOI: 10.26515/rzsi/v124/i1S/2024/172761

[56] Tidemann CR, Nelson JE. Life expectancy, causes of death and movements of the grey-headed flying-fox (Pteropus poliocephalus) inferred from banding. Acta Chiropterologica. 2011;**13**:419–429. DOI: 10.3161/150811011X624901

[57] Wellbergen JA, Meade J, Field HE, Edson D, et al Extreme mobility of the world's largest flying mammals creates key challenges for management and conservation. BMC Biology. 2020;**18**:101+

[58] Hall LS, Richards G. Flying Foxes: Fruit and Blossom Bats of Australia. Malabar FL: Krieger Publishing Company; 2000

[59] Meade J, Martin JM, Welbergen JA. Fast food in the city? Nomadic flying-foxes commute less and hang around for longer in urban areas. Behavioral Ecology. 2021;**32**:1151–1162. DOI: 10.1093/beheco/arab078

[60] Tait J, Perotto-Baldivieso HL, Adam Mckeown A, Westcott D. Are flying-foxes coming to town? Urbanisation of the spectacled flying-fox (Pteropus conspicillatus) in Australia. PLoS ONE. 2014;**9**(10): e109810. DOI: 10.1371/journal.pone.0109810

[61] Boardman WSJ, Roshier D, Reardon T, Burbidge K, et al Spring foraging movements of an urban population of grey-headed flying foxes (Pteropus poliocephalus). Journal of Urban Ecology. 2021;**2021**:1–10

[62] McDonald-Madden E, Schreiber ESG, Forsyth DM, Choquenot D, Clancy TF. Factors affecting Grey-headed Flying-fox (Pteropus poliocephalus: Pteropodidae) foraging in the Melbourne metropolitan area, Australia. Austral Ecology. 2005;**30**:600–608

[63] Nabi G, Wang Y, Lü L, Jiang C, et al Bats and birds as viral reservoirs: A physiological and ecological perspective. Science of the Total Environment. 2021;**754**:142372

[64] Brussow H. From bat to worse: The pivotal role of bats for viral zoonosis. Microbial Biotechnology. 2025;**18**: e70190. DOI: 10.1111/1751-7915.70190

[65] Pauciullo S, Zulian V, La Frazia S, Paci P, Garbuglia AR. Spillover: Mechanisms, genetic barriers, and the role of reservoirs in emerging pathogens. Microorganisms. 2024;**12**:2191. DOI: 10.3390/microorganisms.12112191

[66] White RJ, Razgour O. Emerging zoonotic diseases originating in mammals: A systematic review of effects of anthropogenic land-use change. Mammal Review. 2020;**50**: 336–352. DOI: 10.1111/mam.12201

[67] Mickleburgh S, Waylen K, Racey P. Bats as bushmeat: A global review. Oryx. 2009;**43**:217–234. DOI: 10.1017/s0030605308000938

[68] Mildenstein TL, Tanshi I, Racey PA. Exploitation of bats for bushmeat and medicine. In Voigt CC, Kingston T, editors. Bats in the Anthropocene. Switzerland: Springer International Publishing; 2016. p. 325–376

[69] Jacob ST, Crozier I, William A, Fischer WA, Hewlett A, et al Ebola virus disease. Nature Reviews Disease Primers. 2020;**6**:13

[70] Hassanin A, Nesi N, Marin J, Kadjo B, et al Comparative phylogeography of African fruit bats (Chiroptera, Pteropodidae) provide new insights into the outbreak of Ebola virus disease in West Africa, 2014–2016. Comptes Rendus. Biologies. 2016;**339**: 517–528

[71] Pigott DM, Golding N, Mylne A, Huang Z. Mapping the zoonotic niche of Ebola virus disease in Africa. eLife. 2014;**3**:e04395. DOI: 10.7554/eLife.04395

[72] Olivero J, Fa JE, Real R, Farfán MÁ, et al Mammalian biogeography and the Ebola virus in Africa. Mammal Review. 2017;**47**:24–37

[73] Olivero J, Fa JE, Farfan MA, Marquez AL, et al Human activities link fruit bat presence to Ebola virus disease outbreaks. Mammal Review. 2020;**50**:1–10

[74] Kahn SA, Imtiaz MA, Islam MM, Tanzin AZ, Islam A, Hassan MM. Major bat-borne zoonotic viral epidemics in Asia and Africa: A systematic review and meta-analysis. Veterinary Medical Science. 2022;**8**:1787–1801. DOI: 10.1002/vms3.835

[75] Sánchez CA, Phelps KL, Frank HK, Geldenhuys M. Advances in understanding bat infection dynamics across biological scales. In: Proceedings of the Royal Society B. Vol. **291**. 2024. p. 20232823. DOI: 10.1098/rspb.2023.2823

[76] Peel AJ, Field HE, Aravena MR, Edson D, et al Coronaviruses and Australian bats: A review in the midst of a pandemic. Australian Journal of Zoology. 2021;**67**:346–360

[77] Eby P, Peel AJ, Hoegh A, Madden W, et al Pathogen spillover driven by rapid changes in bat ecology. Nature. 2023;**613**:340–344

[78] Goncalves F, Galetti M, Streicker DG. Management of vampire bats and rabies: A precaution for rewilding projects in the Neotropics. Perspectives in Ecology and Conservation. 2021;**19**:37–42

[79] Rocke T, Streicker DG, Leon A. Management of vampire bats and rabies: Past, present, and future, Vol.1. In Rupprecht CE, editor. History of Rabies in the Americas: From the Pre-Columbian to the Present. Switzerland: Springer Fascinating Life Sciences; 2023. p. 199–222

Chapter 5

Expression of the Enjoyment of Companionship in the Great Horned Owl (*Bubo virginianus*)

Craig T. Perdue and Dwight P. Wynne

Abstract

This is descriptive and exploratory research. Our focus is, first, to build on earlier work by presenting a previously undocumented vocalization and accompanying behavior/context, which most resembles the expression of the enjoyment of companionship in a Great Horned Owl (*Bubo virginianus*). Second, we draw attention to the only other source attributing companionship to *B. virginianus*. Those vocalizations are described qualitatively. We describe putatively similar vocalizations qualitatively and quantitatively from the data of 352 vocalizations we collected and examine their significance as another expression of the enjoyment of companionship. We then place our findings in perspective.

Keywords: companionship, enjoyment, behavior, context, great horned owl, strigids, vocalizations, function, captivity

1. Introduction

Lorenz chose the word "Kumpan" to unify various stimulus-response patterns in geese, as these resemble companionship [1]. The enjoyment of companionship between birds and other animals has been informally discussed extensively for various psittacines as "companion birds." Enjoyment is a mode of consciousness (in the Cartesian sense), and there is evidence for consciousness in non-human animals, including birds, continuing to accrue from neuroscience, including owls [2]. Enjoyment takes states as its objects; we construe states as Chisholm's states of affairs [3]. It has the quality of aboutness, much as propositions or statements do [4]. Moreover, enjoyment may entail a "feeling," over and above pleasure, though what a feeling is and how to explicate it has been a longstanding problem. Whatever feelings are, the consensus view is that various classes of vertebrates and even invertebrates have them [5].

Artuso et al. summarize the work in the field on the types of vocalizations of the Great Horned Owl (*Bubo virginianus*) [6]. However, these vocalizations remain very poorly understood. Where functions are given, Artuso et al. attribute functions to vocalizations based on sources providing non-prospective case studies. The first author (CTP) engaged in over 12 years of collection and examination of primary and

IntechOpen

secondary sources (including video review of infrared remote nest monitoring) to document vocalizations of *Bubo virginianus virginianus* where at least some of the behavioral and contextual information is known. This amounted to 352 recordings from various individuals. Fifty-seven vocalizations given in distinct contexts are from one individual, Rupert, a semi-tame male from the Ypsilanti/Ann Arbor, Michigan area. These recordings span over eight years, from owlet through adulthood. Our focus in this report is, first, to build on the work of Artuso et al. by presenting a previously undocumented vocalization and accompanying behavior/context, which most resemble the enjoyment of companionship. Second, we seek to describe a type of vocalization qualitatively and quantitatively and compare it to qualitative descriptions in Heinrich, the only other source attributing companionship to *B. virginianus virginianus* [7]. (There is no documentation for other subspecies.) We then put these in perspective.

2. Methods and Discussion

CTP recorded all of the vocalizations of interest in the field. The previously undocumented vocalizations were recorded opportunistically as. wav files with an Olympus Portable Hand-held Recorder DM-620, held approximately 20 centimeters away. The rest of the vocalizations were recorded using a Marantz PMD-661Mkiii Field Recorder with a Sennheiser ME 66 Shotgun Microphone. As available, recordings were made in 19 distinct types of contexts.

Rupert came to the first author at approximately 4 weeks of age, raised by his parents. He was developed as a bird of falconry. Training, using only positive reinforcement, began within days of being introduced to his new home. CTP spent time daily with Rupert, including visitation in his indoor enclosure, i.e., a mew, which was a separate converted bedroom.

Spectrograms, waveforms, and their associated quantitative measurements were extracted using Raven Pro 1.5.0 build 43. The following parameters were used: Window Type: Hann, Window Size: 512, DFT Size: 1024, Overlap: 90. Frequency was measured to a lowest limit of 60 Hz for two reasons. 60 Hz is the lowest frequency at which *B. virginianus* can hear [8], and there was a band of recording noise located below 60 Hz. We analyzed vocalizations using the following features: frequency (first and third quartile), duration, sound intensity (classified by CTP as low, medium, or high intensity relative to the other vocalizations), and the number of syllables.

The previously undocumented type of vocalization was recorded by CTP outside Ypsilanti, Michigan, in a vacant middle school parking lot at approximately 21:30 (EST), 16 October 2014, on a dark night. At the time of recording, he was approximately 8.5 months old. Rupert was recorded shortly after a free-flight exercise in a field adjacent to the school with a lit parking lot. He landed on asphalt, at the base of a lit light pole approximately six meters away from the first author. Rupert used a rapid side-to-side gait for 2 meters to get from a nearby flagpole to where the first author was seated on the pavement. He then situated himself in the author's lap in a prone position across the author's lap. He faced toward the school situated approximately 35 meters away. No other events occurred in the surrounding environment.

In this context, Rupert transmitted at least three faint pulses, with each pulse separated by approximately five seconds. A pulse has been described elsewhere for anurans as a single burst of sound energy, not further subdivided into subunits, separated by strong amplitude modulation from other pulses [9]. The pulses were recorded opportunistically as a result of the accompanying unique non-vocal behavior, which strongly resembled the enjoyment of companionship. They could theoretically be the result of chaos or randomness; however, they occurred in a patterned series, and their functional role as expressing the enjoyment of companionship is *prima facie* evident (in exploratory research) given behavior and context. Further, the contextual and behavioral characteristics point to the occurrence of the enjoyment of companionship as the inference to the best explanation since, as far as we have been able to ascertain, no other explanation renders them less surprising [10]. For example, it is harder to see how a case for affection is as strong. Because *B. virginianus* is an intrinsically aggressive, solitary animal, feeling affectionate toward another animal is therefore not likely *a priori*. Upon review of allopreening in this species in the literature, on the Internet, and in the first author's video archives, allopreening occurs between parent and offspring in the nest and among branchers. There was one instance in adults of what presumably was a pair. Kenny et al. describe a prevalent pair bond maintenance function for allopreening in birds [11]. However, pair bonding does not entail affection. Because of prey-strewn nest conditions for this species, allopreening between parent and young may just as well play a hygienic (anti-parasite) function between parent and young. Because *B. virginianus* is occasionally cannibalistic between adults and there are crowded conditions among branchers, allopreening among branchers may perform a stress reduction strategy or provide heat conservation. The case described here – clearly not possible in virtue of these explanations – is more likely to obtain because it expresses a positive disposition toward the *state* of the agents being together.

The pulses recorded – vocalizations 348 and 349 – have a low frequency (ranging from Q1 = approximately 200 Hz to Q3 = approximately 600 Hz) and a duration of 30–40 ms, at least four times briefer than any of the other 352 vocalizations. Both had an extremely low acoustic intensity, audible to the human ear only from a very close distance. **Figure 1** presents the spectrogram of 348, which closely resembles the other pulse recorded, vocalization 349.

If CTP and Rupert were companions, it is reasonable to have observed the behavior and vocalizations more than once. That is not the case here. At the time the behavior was observed, Rupert was a young bird (about 8.5 months old). At some point subsequent to the pulses being recorded, the then relatively ignorant first author engaged Rupert in negative interactions over a period of several months in the process of forcing radio telemetry to be affixed. (Damage done has since been only

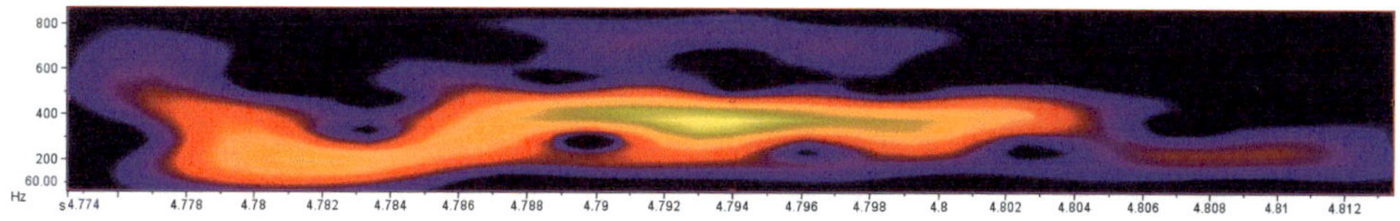

Figure 1.
Representative pulse of subject (vocalization 348).

recently overcome after a process of years.) It is possible that the pulses reflect Rupert's developmental stage or these expressions reveal an intimacy that was lost due to the negative interactions.

There is only one other source where the enjoyment of companionship has been stated for *B. virginianus*. Heinrich coexisted in the same dwelling with his bird for years, and the bird was subject to minimal restrictions. Heinrich explicitly identifies one type of vocalization as expressing the enjoyment of companionship, or what he simply calls "companionship" [7].

According to Heinrich, "I hear them [vocalizations] only when I am next to him; they are his private sounds, reserved for intimacies.... It is these ...that bond friendship." They are "...quick little nasal, reedy chuckles varying in tone".

Vocalization 16, also from Rupert, was the most audible of three similar vocalizations of low intensity in the session during which it was recorded, which occurred approximately six months after the pulses (and well before the traumatic events described). These three vocalizations match the qualitative description given by Heinrich. None other of the 352 fit this description. Like 348 and 349, 16 was faint. Because of this and a noisier environment, its spectrogram is not perspicuous (**Figure 2**).

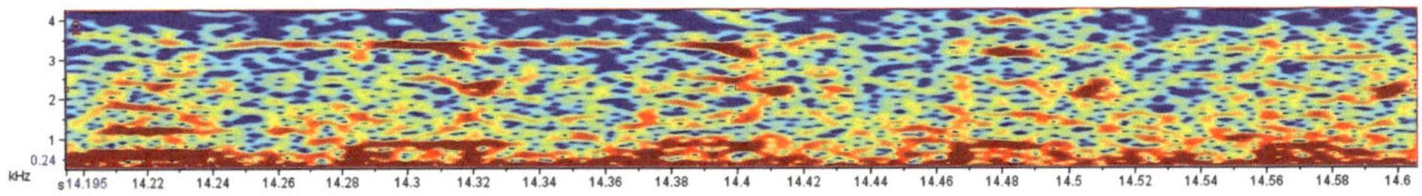

Figure 2.
Qualitatively matching vocalization 16.

Though 16 had five syllables, it was still quite brief, lasting 0.41 seconds, with a syllable rate of approximately 12.2 syllables per second. (16) has a frequency spanning approximately 400–650 Hz, similar to that of vocalizations 348 and 349. Both contribute to a similarity in this respect (**Figure 3**). The intensity level is the same as well.

To quantify how different the pulses were from the other vocalizations recorded from Rupert, we developed a metric representing the "distance" between any two vocalizations. We log-transformed the duration, number of syllables, and Q1 and Q3 frequency, and standardized each transformed variable to have mean 0 and standard deviation 1. For the intensity variable, we represented low intensity as -0.5, medium as 0, and high as 0.5. Finally, we computed the Euclidean distance between the vocalizations based on these transformed and standardized variables. This analysis was performed using R version 4.1.1 [12]; graphics were produced using the package ggplot2 version 3.3.6 [13].

The minimum of the two distances to 348 and 349 was considered as the "overall distance" between a vocalization and the pulses. As shown in **Figure 4**, no other vocalization is closer to the pulses than 16.

The context/behavior of 16 involved Rupert standing stationary on the ground in an area where Rupert had been trained before using strictly positive reinforcement. He stood stationary between two women who were lying and sitting on the ground, flanking Rupert opposite one another. Each was approximately 6 meters from the

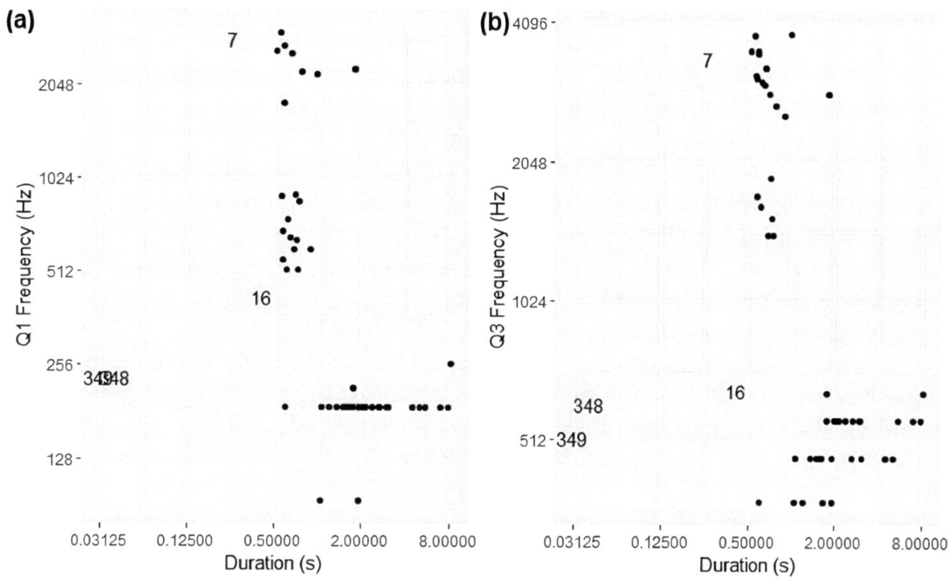

Figure 3.
Frequency vs. Duration for all of the subject's vocalizations.

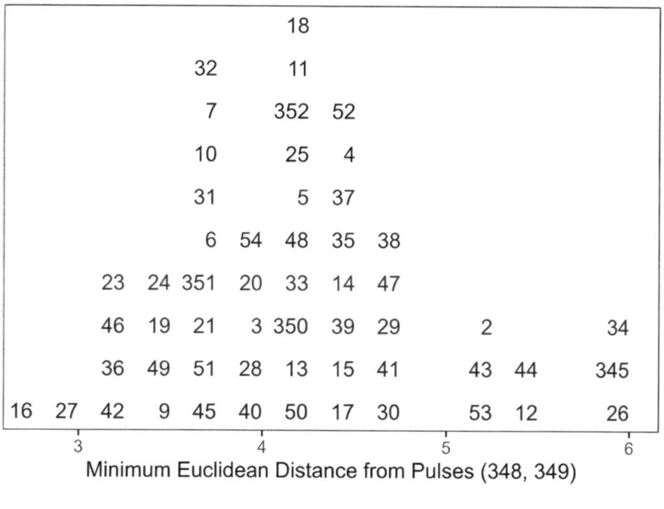

Figure 4.
Minimum (Euclidean) distances from pulses (348,349).

bird. The women softly praised the bird in high, soft voices each time the bird vocalized, and soft encouragements preceded these vocalizations. (16) and its cohorts were the only vocalizations that occurred. Though there are other explanations for 16 with its accompanying behavior/context, the expression of the enjoyment of companionship is also compatible.

For vocalization 16, its occurrences need to be further elucidated. From Heinrich's work [7], vocalizations which resemble it appear to express the enjoyment of companionship. For the pulses, their associated behavior and context certainly point to that dispositional attitude. Perhaps Rupert and Heinrich's owl are unique, anomalous cases. That, at least, appears doubtful to us. However, corroborating evidence is necessary. That may be attainable because both birds exhibited their behavior with allospecifics in captivity, and because the occurrence of this species is widespread in the network of zoos and rehabilitation facilities.

The challenge of the reproducibility of our findings is daunting because documentation for the development of antecedent conditions does not exist for the Great Horned Owl. Though it is anecdotal, it is beyond speculation that CTP contends that this can be done with sufficient detail. There are also a number of intriguing testable hypotheses related to the similarity of the acoustic properties described here and commonalities in functional roles. Of course, there are questions about how to provide the best welfare for individuals kept in captivity (perhaps including other strigids). As an intrinsic good on hedonistic grounds and some other moral theories, their enjoyment merits the effort [14].

Acknowledgements

Above all, we want to thank the City of Ann Arbor's Natural Area Preservation division for their support. We also thank Jorg Massen, Fred Feldman, and Susan Volman for their valuable insights. Thanks are due to the numerous contributors of data and to those facilities that allowed us to collect data from their captive birds. Special thanks to Michael Pitzrick, Yang Center for Conservation Bioacoustics, Cornell University, for his tireless support and generosity.

There are no competing interests to declare.

The authors confirm that the research meets all required ethical guidelines, including adherence to the legal requirements of the United States Fish and Wildlife Service and the United States Department of Agriculture Animal Welfare Act standards for birds.

The data constitute part of an ongoing project, and it will be publicly available once this project comes to an end.

Author details

Craig T. Perdue[1]* and Dwight P. Wynne[2]

1 Shadow Speak, LLC, Ypsilanti, MI, USA

2 Department of Mathematics, California State University, Fullerton, CA, USA

*Address all correspondence to: cperdue@shadowspeak.org

IntechOpen

© 2026 The Author(s). Licensee IntechOpen. This chapter is distributed under the terms of the Creative Commons Attribution License (http://creativecommons.org/licenses/by/4.0/), which permits unrestricted use, distribution, and reproduction in any medium, provided the original work is properly cited.

References

[1] Lorenz K. The Companion in the bird's world. The Auk. 1937;**54**(3): 245–273. DOI: 10.2307/4078077

[2] Stacho M, Herold C, Rook N, Wagner H, Axer M, Amunts K, Gunturkun O. A cortex-like canonical circuit in the avian forebrain. Science. 2020;**369**(65). DOI: 10.1126/science. abc5534

[3] Kim J. States of affairs, events, and propositions. Grazer Philosophische Studien. 1979;7:147–162

[4] Feldman F. The good life: A defense of attitudinal hedonism. Philosophy and Phenomenological Research. 2002;**65** (3):604–628

[5] de Waal F, Andrews K. The question of animal emotions. Science. 2022;**375** (6587):1351–1352. DOI: 10.1126/science. abo2378

[6] Artuso C, Houston C, Smith D, Rohner C. Great Horned Owl(Bubo virginianus), Version 1.1. Birds of the World Ithaca. Sly ND, Editor. Cornell Lab of Ornithology; 2022. DOI: 10.2173/ bow.grhowl.01.1

[7] Heinrich B. One Man's Owl. Princeton: Princeton University Press; 1987

[8] Edwards E. Hearing ranges of four species of birds. The Auk. 1943; **60**:239–241

[9] Köhler J, Jansen M, Rodríguez A, Philippe J, Toledo L, Emmrich M, Glaw F, Haddad C, Rödel M, Vences M. The use of bioacoustics in anuran taxonomy: Theory, terminology, methods and recommendations for best practice. Zootaxa. 2017;**4251**. DOI: 10.11646/ zootaxa.4251.1.1

[10] Schupbach J, Sprenger J. The logic of explanatory power. Philosophy of Science. 2011;**78**(1):105–127. DOI: 10.1086/658111

[11] Kenny E, Birkhead T, Green J. Allopreening in birds is associated with parental cooperation over offspring care and stable pair bonds across years. Behavioral Ecology. 2017;**28** (4):1142–1148. DOI: 10.1093/beheco/ arx078

[12] R Core Team. R: A language and environment for statistical computing. R: Foundation for Statistical Computing. 2021. Available from: https://www.R-project.org [Accessed: 2026-February-24]

[13] Wickham H. Ggplot2: Elegant Graphics for Data Analysis. Berlin: Springer-Verlag; 2016

[14] Feldman F. What Is This Thing Called Happiness? Oxford: Oxford University Press; 2010